Einstein Was Wrong!

But So Was Newton

I0476582

Introducing
The Galaxy Gravity Cycle (GGC)
The Atomic Model of Motion (AMM)

Martin O. Cook

Parts Edited by Stephen Gibbons

Insightful Contributions by David Miller, Monte Mower, Jeffery Davis, **Wendy Cook**, Stephen Gibbons, Jack Hylton Jr.,Bryan McPherson, Randy Dean, Mike Scott, John Cook Sr., Sally Cook, Kylene Schramm, Thomas Cook, Sarah Cook, Rebekah Cook, Isaiah Cook, Kathryn Cook, and Kelsey Dean Grace Cook

Copyright © 2015 Martin O. Cook
All rights reserved.
ISBN-13: 978-1515022909
ISBN-10: 1515022900

Modern Physics is Built on an Erroneous Foundation

The purpose of this book is to expose the biggest error in scientific history and introduce two theories to rectify this error. This crucial misperception has its roots in the scientific ideas of Galileo and Newton. In 1905, Einstein's special theory of relativity perpetuated the same false premise held by Galileo and Newton. Then in 1915, Einstein's general theory of relativity solidified the endurance of this error for at least another century. In essence, we could say that Galileo and Newton started the great misperception; Einstein perpetuated it, and scientists today are still riding on its wake.

Why is this do important? Even though Newton lived over 300 years ago, scientists, to this day, still do not understand how gravity is sustained from an atomic/quantum perspective. This is a direct result of the great misperception made by Galileo and Newton and perpetuated by Einstein. Even Einstein, who is considered by many as the greatest scientist of all time, failed in his attempts to unify gravity with atomic/quantum processes.

> "For decades Einstein attempted to develop a unified field theory...connecting the movement of planets and stars with the operations of the tiniest subatomic particles." (Lacayo, 2014, 9)

What Albert Einstein didn't realize is that his two relativity theories did more to perpetuate the gap between our present understanding of gravity and the atomic/quantum processes involved in sustaining gravity than any other theory or collection of theories.

The first main obstacle perpetuating the gap between our present understanding of gravity and the atomic/quantum processes involved in sustaining gravity is the great misperception that was fathered and sustained by Galileo, Newton, and Einstein. The first part of this book will address the great misperception.

The second main obstacle perpetuating the gap between our present understanding of gravity and the atomic/quantum processes involved in sustaining gravity is the lack of an atomic/quantum model that explains the momentum, relativity, and gravity of masses from an atomic/quantum perspective. With blind acceptance of the great misperception strangling the creativity of present day physicists, there has been no need to attempt to explain how atoms move through space in the first place, the very foundation for understanding the atomic/quantum processes involved in sustaining gravity. The second and third parts of this book will bridge the gap between our present understanding of gravity and the atomic/quantum processes involved in sustaining gravity by introducing two new theories: The Atomic Model of Motion (AMM) and The Galaxy Gravity Cycle (GGC).

Before I begin a childlike explanation of the Galaxy Gravity Cycle (GGC), the overarching big picture of this book, I want to ask and attempt to answer one simple question. Why is gravity so important?

Gravity is the glue of the universe. It holds galaxies and solar systems together while keeping people and things bound to the Earth. It is acting on us every second of every day during our entire life. Without gravity, we fall off the planet while moons, planets, and stars fall out of orbit. Gravity is the unifying mechanism of the universe—the unifying mechanism of every galaxy, of every solar system, and of every planet. Gravity is the unifying mechanism upon which all of life and physics have evolved.

The Galaxy Gravity Cycle (GGC)—A Childlike Explanation

How do we know when we really understand something? We see how it operates within a cycle. For example, we better understand evaporation and precipitation when we see how they operate within the bigger picture of the water cycle. The same is true for motion and gravity. We better understand them when we see how they operate within the bigger picture of the Galaxy Gravity Cycle (GGC). The Galaxy Gravity Cycle is a repeating pattern that keeps the flow of energy (gravity) moving through a galaxy. The Galaxy Gravity Cycle, like the water cycle, can be broken down into steps in order to easily explain the overall picture of how it works. Like the water cycle, the Galaxy Gravity Cycle is intuitively easy to understand.

Step 1: Energy continuously flows out from the center of a galaxy. Stars already in motion absorb this energy. This energy is the gravity that keeps a star's motion in an orbital pattern around the center of a galaxy.

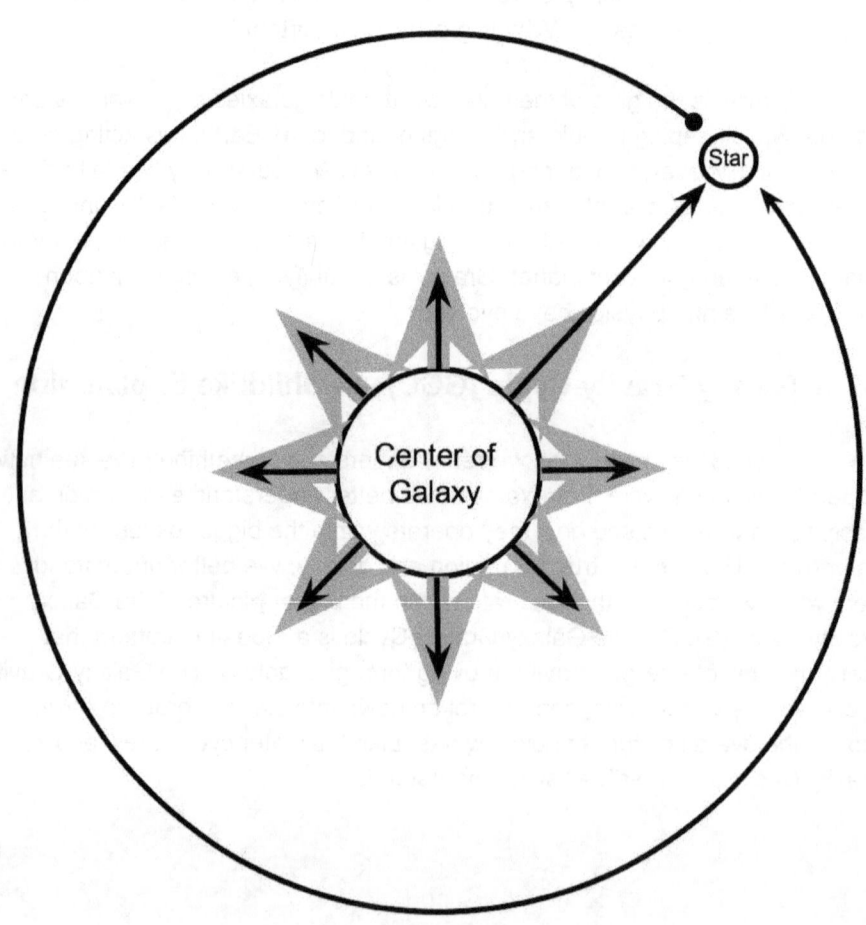

Step 2: As a star continuously absorbs energy, it also continuously emits energy. It does so proportionately through its spherical surface. This causes planets and other objects already in motion to orbit around its spherical shaped body as it orbits around the center of the galaxy.

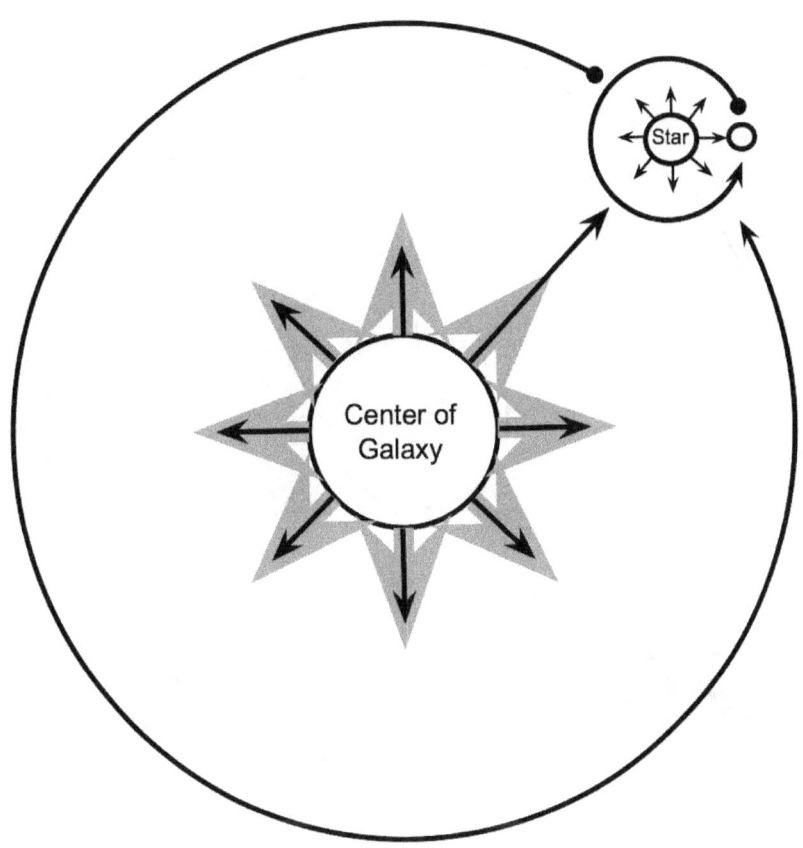

Step 3: Planets that are already in motion continuously absorb energy coming from the star they orbit. This energy is part of the gravity process that keeps them orbiting the star. As planets absorb energy, they also emit a steady flow of energy through their spherical surfaces. This causes other objects like a moon (that is already in motion) to orbit the planet. The emitted energy also keeps objects like balls, cars, bikes, humans, and animals that are already in motion from floating off into space.

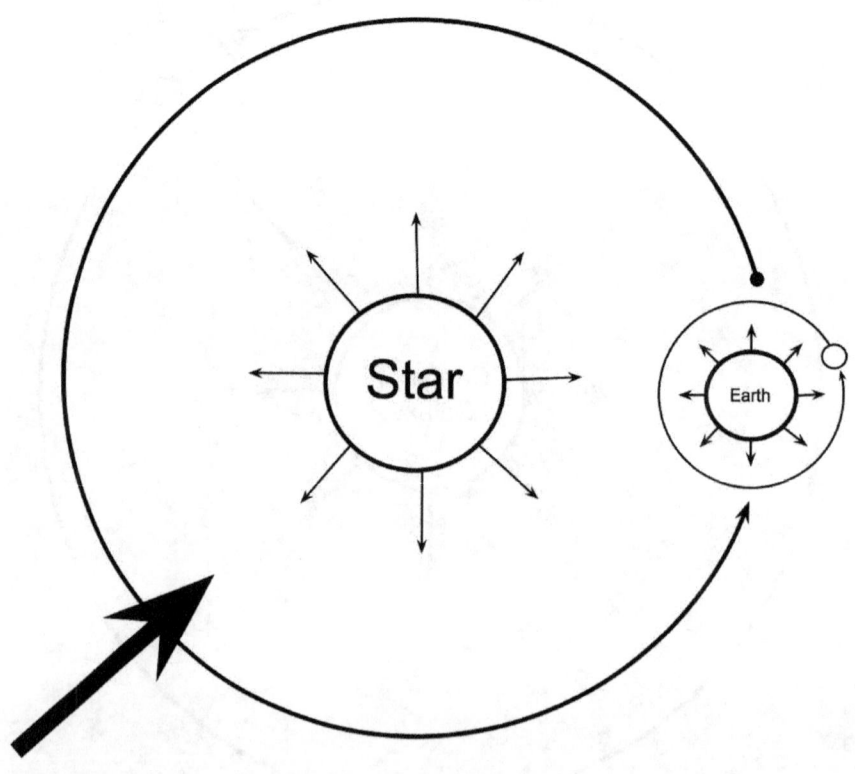

Step 4: Much of the energy that is emitted by the center of the galaxy makes it back to the center of the galaxy, where it is absorbed and redirected back out into the galaxy.

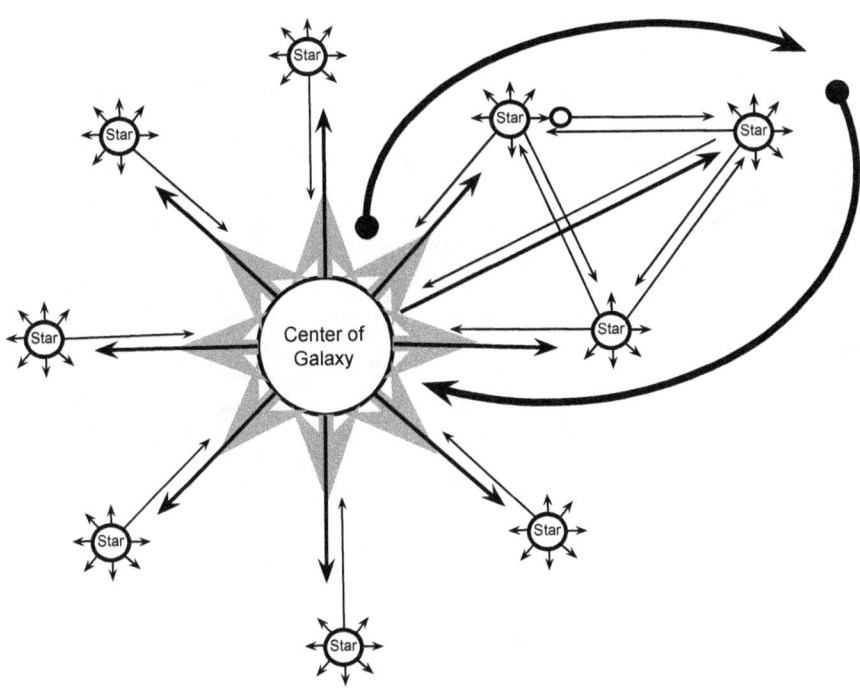

A complex web of spherical bodies sustains the energy cycle within each galaxy.

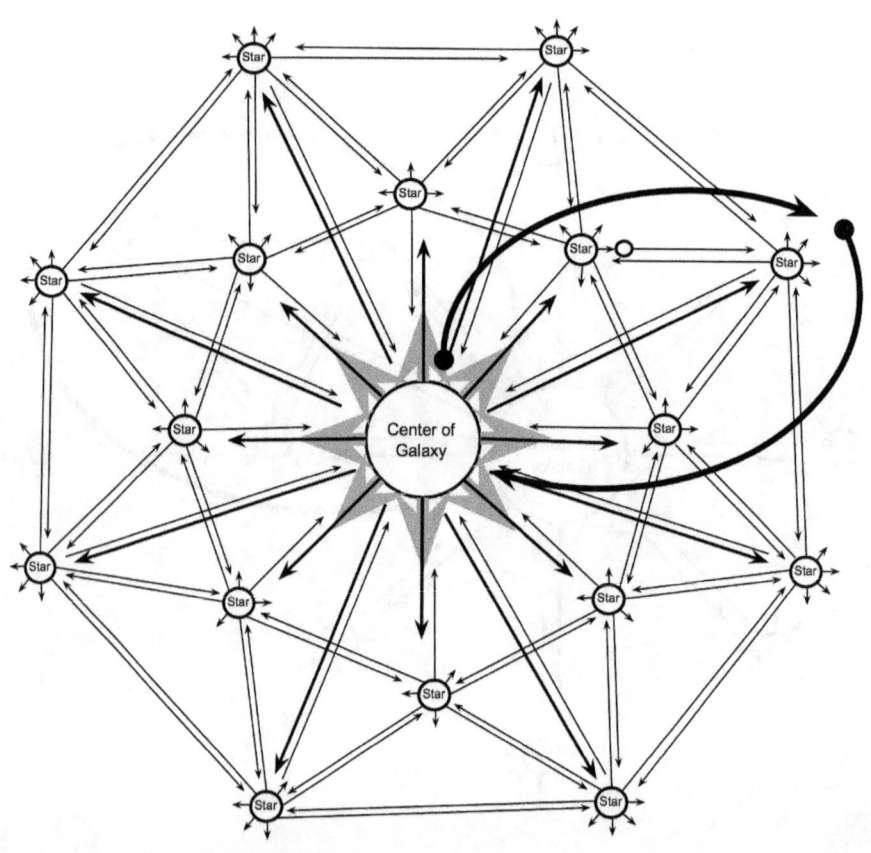

Step 5: Energy that escapes the energy cycle of a galaxy feeds into other galaxies. In turn, galaxies receive energy from other galaxies to replace the energy they have lost. Our galaxy receives lost energy from other galaxies to replace its lost energy. Galaxies, like stars, play an important role in keeping the balance of energy flowing throughout the universe.

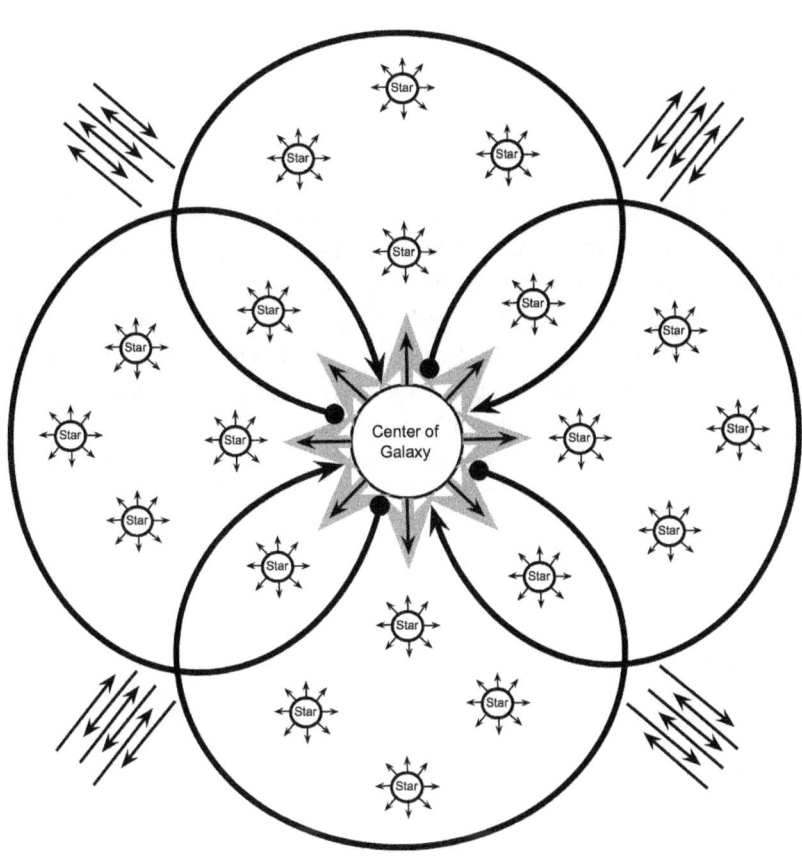

In summation, the cyclical flow of energy within a galaxy is the gravity that holds that galaxy together. Spherical bodies are the vehicles through which this energy is absorbed and then redirected in all directions. As with any cycle, the process that keeps the energy flowing within a galaxy repeats itself over and over and over again. And as some energy will escape the gravity cycle of our galaxy, our galaxy absorbs energy that has escaped the gravity cycle of other galaxies. The Galaxy Gravity Cycle (GGC) is the cyclic process that holds all moving bodies in orbit and keeps humans from floating off into space.

Overview

The remainder of this book is divided into three parts. The summation of these three parts fully addresses the two main obstacles perpetuating the gap between our present, limited understanding of gravity and the atomic/quantum processes involved in sustaining gravity. Part One focuses on **the great misperception** fathered and sustained by Galileo, Newton, and Einstein. The great misperception is a very limited view of how and why objects move through space, limiting our understanding of the momentum, relativity, and gravity of atoms and masses. Part Two focuses on the **Atomic Model of Motion (AMM)**. The atomic model of motion explores the atomic/quantum processes involved in the momentum, relativity, and gravity of atoms and masses. Part Three focuses on the **Galaxy Gravity Cycle (GGC)**. The Galaxy Gravity Cycle is the big picture that explains the flow of energy within a galaxy needed to sustain the momentum, relativity, and gravity of atoms and masses. This flow of energy holds objects in orbit and prevents people and things from floating off into space. These three topics make up the three parts of this book.

1. The Great Misperception (**Part I**)
2. The Atomic Model of Motion (**Part II**)
3. The Galaxy Gravity Cycle (**Part III**)

Part I
—The Great Misperception—
—From Newton's Mistake to Einstein's Real Blunder—

Definitions:

The Great Misperception: The great misperception is the mistaken assumption that an object at rest tends to stay at rest. (There is no such condition as an object at rest.) This misperception has its roots in Galilean relativity when Galileo recognized that he could not tell the relative motion of his boat to the land when applying the laws of physics within a confined cabin on the boat. Newton cements this misperception by stating in his first law of motion that an object at rest tends to stay at rest. According to Galileo and Newton, objects at rest within any inertial frame are getting a free ride. This misperception gave birth to the motion myth.

The Motion Myth: The motion myth is the mistaken assumption that masses just mysteriously move through space without any atomic/quantum accountability justifying motion and changes to motion.

Newton's Mistake

How or why do masses move through space in the first place? The colossal failure to at least attempt to answer this question of how masses move through space in the first place is the biggest error in physics. It is the very reason that an understanding of how atomic/quantum processes are involved in sustaining gravity still remains a mystery. Isn't it ironic that the very person who coined gravity, who probably did more for physics than any one individual except for maybe Albert Einstein, impeded its atomic/quantum explanation for centuries?

What was Newton's Mistake? Newton didn't acknowledge that all objects are always in motion, even when they appear to be at rest. He viewed objects, whether in motion or at rest, from their whole perspective and not from the individual parts (atoms) that form each object. In doing so, he cemented the precedence that masses just mysteriously move through space or can just take a position of rest within an inertial frame. This mindset prevented him from hypothesizing a theory to try to explain the science of motion to coincide with his laws of motion. Such a theory would have attempted to explain how objects move through space in the first place, potentially exposing the motion myth

14

(that objects just mysteriously move through space without any atomic/quantum accountability) at the dawn of the scientific age. This same misperception tainted Einstein when he was putting forth his relativity theories. Einstein's theories do well to predict results but fail to incorporate atomic/quantum processes. Modern physicists still lack a viable theory to explain from a quantum perspective how and why masses move through space in the first place.

So why blame Newton? If Galileo before him and Einstein after him are guilty of making the same mistake, then why refer to this colossal failure as Newton's Mistake? The thesis of this book is that this mistake, (whether made by Galileo, Newton, Einstein, or physicists today), is so crucial that it plagues our ability to explain how gravity really works. So a better question may be: Why not blame Newton? Newton authored the laws of motion. If the laws of motion are incomplete, if they don't tell the whole story about motion, should we not go back to the source? Newton's mistake perpetuated the motion myth and is the very reason Einstein was wrong.

Newton said he could see further than others because he stood on the shoulders of giants. What about his shoulders? Who is standing on Newton's shoulders? Essentially, everyone who came after him is standing on his shoulders. That is why his mistake has had a blinding effect for centuries.

Why is exposing Newton's mistake so crucial to understanding how gravity really works? Simply put, gravity cannot be reconciled with atomic/quantum processes without an atomic model of motion to tie together our present understanding of gravity with the atomic/quantum processes involved in sustaining gravity. An atomic model of motion explains how masses move through space in the first place. From there, one can easily surmise how gravity really works. Since an atomic model of motion didn't cross Newton's mind, it somehow also slipped past the minds of all those who stood on his shoulders. By exposing Newton's mistake, we get to the source of the problem quicker.

Even though Newton's mistake has carried over to our day, I have no problem giving everyone a pass up to the inception of quantum mechanics. Even Einstein deserves a pass since he developed his relativity theories just prior to the formation of quantum mechanics. (He did spend the declining years of his life, after the inception of quantum mechanics, searching for a unification theory, but to no avail.) Current scientists face a different dilemma. Quantum physics has been around for a while. We have insights that weren't available to pre-quantum scientists. So why aren't we any closer to unifying our present understanding of gravity with atomic/quantum processes? The reason is because we are still stuck in the quagmire of Newton's mistake. The only way our present understanding of gravity and atomic/quantum processes will ever be reconciled is with an atomic model of motion to tie them together. An atomic model of motion unites the differences between Newton's laws, Einstein's relativities, and atomic/quantum processes into a single, viable theory. All three of these seemingly separate laws and theories are a part of the same atomic/quantum processes. An atomic model of motion will explain atomic/quantum processes involved in the motion of objects through space, including the acceleration we call gravity. In order to explain how an atomic model of motion works, we first have to untangle the knot caused by Newton's mistake.

Newton's mistake ushered in a crucial misconception. The very person credited with discovering gravity hindered its ultimate explanation when he declared that a body at rest tends to stay at rest. For Newton, bodies were either in motion or at rest, which is understandable from his earthbound perspective. But from an atomic/quantum perspective, masses are never at rest; all masses are always moving though space. Even objects that appear to be at rest are still moving through space. What is important to realize is that it is the individual atoms that make up masses that are responsible for this perpetual movement through space. (This will be explained in detail in part two of this book.) This perpetual motion of all masses is *the great secret of physics* that links classical physics to atomic/quantum processes and is the key to understanding how gravity really works.

Think of astronauts experiencing a continual free fall as the space shuttle orbits the earth. As the astronauts appear to be floating within the confines of the shuttle, one astronaut passes an orange to another astronaut. As the orange makes the trek through space, its motion is visible by both astronauts. When the astronaut receiving the gift grabs it and places it in space next to him, the orange now appears to have stopped moving as it takes a position of rest next to the astronaut who placed it there. Is the orange really at rest? What appears to be at rest to the astronaut who just placed it in space next to him is really freefalling around the earth at 17,500 miles per hour. The orange is still in motion through space. Although objects can appear to be at rest, it doesn't mean they are not moving through space. They still have momentum. This is what Newton failed to grasp when he saw objects at rest on the earth's surface and categorized them differently than objects in motion. The objects that appear to be at rest are not getting a free ride. (The concept of not getting a free ride will be explained in greater detail in part two of this book.)

Galileo also failed to visualize that all objects have perpetual momentum. If confined in a windowless cabin of a uniformly moving ship, Galileo said he wouldn't be able to tell the ship's speed relative to the land by applying what he knew about the laws of physics. He rightly reasoned that the laws of physics are the same for all inertial frames. All objects that appear to obtain a state of rest within any inertial frame mysteriously abide by the same laws of physics. In any inertial frame, if you throw a ball straight up, it will come straight down. If you put the same ball in a different inertial frame and throw it straight up, it will again come straight down. What he saw with his eyes was correct, but what he didn't see tells the rest of the story. Both Newton and Galileo were unaware that objects that appear to be at rest are never truly at rest. They are always in motion. As I will detail later, the collection of atoms making up any mass are the source and cause of the momentum of that mass through space. Due to their limited perspective, Newton and Galileo were completely oblivious to atomic/quantum processes operating within all masses. The atoms of masses continuously drive the perpetual momentum of these masses, even if they appear to be at rest within an inertial frame, such as the inertial frame of Galileo's ship or the inertial frame of an object resting on the surface of the earth.

What makes the perpetual momentum of all masses so indiscernible is that bodies at rest on the earth's surface really appear to be at rest. Their motionless bodies could not possibly be in motion and maintain that motion until an uneven force acts upon them. This is the vantage point from which Newton saw objects at rest. Like Newton, our minds do not naturally perceive that objects that appear to be at rest are actually in motion through space and the cause of that motion being the individual atoms that make up that object. The motion of objects that appear to be resting on the earth is tied to their gravity-bound status to the earth, sharing the same motion as the earth, similar to a comb sitting on a seat of a moving car sharing the same motion as the moving car. Once you slam on the breaks, it becomes apparent that the comb's motion is independent of the motion of the seat upon which it appeared to be resting. This misperception that objects can obtain a state of rest has been passed down to us and continues to veil how atomic/quantum processes are involved in sustaining gravity.

In reality, you cannot separate the physics of bodies in motion from the physics of bodies that appear to be at rest. All objects are always in motion even when they appear to be at rest. To emphasize this point, imagine a switch that turned off the gravity that accelerates objects into the earth's surface. With the effects of gravity shut off, the slightest nudge to any of these objects would cause these objects to move away from the earth's surface. The nudge wouldn't begin the motion; it would only initiate a slight variance of the motion they already exhibited when they were moving through space attached to the hip of the earth by the effects of gravity. As these nudged objects, once bound to the earth's surface, continued in their new momentum path, moving away from the earth's surface, one would witness that these objects always had motion. Their perpetual motion was cloaked by their gravity-bound status to the earth's surface.

Newton's mistake was his failure to perceive that all bodies or objects are always in perpetual motion through space and are never at rest. Within this perpetual motion, they can either accelerate or decelerate but at no time do they ever stop being in motion through space. Newton could have declared that all bodies are always in motion and are never at rest even when they appear to be at rest, and that all objects tend to maintain the same motion unless acted upon by an uneven force. This would have corrected a great misperception that has stifled an atomic/quantum explanation of gravity up to this point in human history. Perpetual momentum is the natural state of all masses and acceleration and deceleration are only temporary alterations to the equilibrium of perpetual momentum. The reality that all objects are always in motion is the *great secret of physics* that was firmly veiled by Newton's mistake.

Had Newton tried to devise a theory for the science of motion, he might have recognized his own mistake or at least planted the seeds that would inspire those standing on his shoulders to ask: How do objects move through space in the first place? In other words, what is happening on an atomic/quantum level that allows a body in motion to stay in motion? Or even more basic, what is happening on an atomic/quantum level that allows a body, any body, to be in motion at all or have any movement whatsoever? In short, what is the role of atomic/quantum processes in the momentum and movement of mass through space? By failing to hypothesize why masses move through space in the first place, a faulty paradigm was adopted into the scientific community, implying that masses, which are made up of atoms, just mysteriously move through space. Newton's mistake perpetuated the motion myth.

The Motion Myth

During a crucial part of a game, a coach will call a timeout to assess the situation and prepare his players to execute at the highest level, desiring the greatest outcome possible. Presently, in theoretical physics, a timeout is needed to assess the direction in which it is headed. Even with the discovery of the Higgs boson, the Standard Model cannot explain how gravity really works.

The failure to grasp that all objects are always in motion, even when they appear to be at rest, continues to erect an impenetrable barrier that prevents modern physicists from discovering the atomic/quantum processes involved in the momentum, relativity, and gravity of atoms and masses. In order to rectify hundreds of years of tradition, we must correct Newton's mistake at its origin and then move forward from there.

The following illustrates consequence of Newton's mistake in perpetuating a paradigm that leads to a dead end.

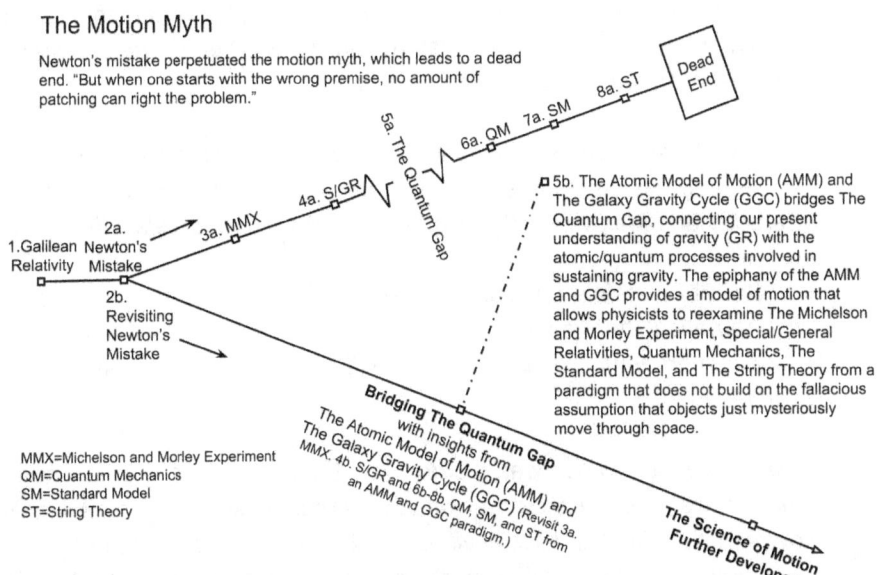

The Motion Myth

Newton's mistake perpetuated the motion myth, which leads to a dead end. "But when one starts with the wrong premise, no amount of patching can right the problem."

1.Galilean Newton's Relativity Mistake
2a.
2b. Revisiting Newton's Mistake
3a. MMX
4a. S/GR
5a. The Quantum Gap
6a. QM
7a. SM
8a. ST
Dead End

5b. The Atomic Model of Motion (AMM) and The Galaxy Gravity Cycle (GGC) bridges The Quantum Gap, connecting our present understanding of gravity (GR) with the atomic/quantum processes involved in sustaining gravity. The epiphany of the AMM and GGC provides a model of motion that allows physicists to reexamine The Michelson and Morley Experiment, Special/General Relativities, Quantum Mechanics, The Standard Model, and The String Theory from a paradigm that does not build on the fallacious assumption that objects just mysteriously move through space.

Bridging The Quantum Gap with insights from The Atomic Model of Motion (AMM) and The Galaxy Gravity Cycle (GGC) (Revisit 3a. MMX. 4b. S/GR and 6b-8b. QM, SM, and ST from an AMM and GGC paradigm.)

MMX=Michelson and Morley Experiment
QM=Quantum Mechanics
SM=Standard Model
ST=String Theory

The Science of Motion Further Developed

19

Path (a) represents the path we are currently traveling down. Any theory based on a motion myth paradigm will eventually lead to a dead end.

Path (b) represents a new path that will lead to the further development of the science of motion.
The following gives a brief description for the different points along the paths.

1. (Galilean Relativity) The Motion Myth begins with Galileo's relativity observations when he realizes that the same object in different inertial frames follows the same laws of physics. Galileo fails to question the source and cause of this phenomenon from an atomic/quantum perspective. When quantum physics comes onto the scene, it is not invoked as potentially being able to explain Galilean relativity from an atomic/quantum perspective.

2a. (Newton's Mistake) Newton misperceives that objects can obtain a position of rest. From an atomic/quantum perspective, there are no free rides. All objects are always in motion, even objects that appear to be at rest. Newton's mistake perpetuated the motion myth. 2b. By correcting Newton's mistake at its origin, we eradicate the motion myth paradigm. We begin a new path that acknowledges that all objects are always in motion, even when they appear to be at rest—the first step in the development of the science of motion.

3. (MMX) The Michelson and Morley Experiment played a pivotal role in the progression of physics. The 1887 experiment follows the work of Faraday and Maxwell and precedes Einstein's special relativity.

4. (S/GR) Einstein's Special and General Relativities are results-based mathematical theories that make accurate predictions but fail to explain momentum, relativity, and gravity from an atomic/quantum perspective.

5a. The quantum gap was created with the ushering in of quantum mechanics. Quantum mechanics could not be linked to Newton's laws of motion or Einstein's Relativities. Each of these three different viewpoints requires a different mindset to understand. 5b. The Atomic Model of Motion (AMM) and the Galaxy Gravity Cycle (GGC) are process-based theories that tie together Newton's laws, Einstein relativities, and atomic/quantum physics.

6. (QM) Quantum Mechanics deals with the mathematical description of the motion and interaction of subatomic particles but fails to explain why masses as a whole move through space in the first place. Consequently, it fails to explain the atomic/quantum processes involved in the momentum, relativity, and gravity of masses. This failure directly stems from Newton's mistake.

7. and 8. (SM and ST) The Standard Model and String Theory are byproducts of a motion myth paradigm. "But when one starts with the wrong premise, no amount of patching can right the problem." (Seifer, 1996, 21)

The Motion Myth remains the main obstacle preventing us from understanding how momentum, relativity, and gravity operate from atomic/quantum processes.

Einstein's Real Blunder

[Time dilation and space-time are products of the motion myth paradigm Einstein was stuck in when he developed his Relativity theories. They are mathematical substitutes for atomic/quantum processes.]

One could scour through the mathematical formulas establishing Einstein's relativity theories and not see the error of his way. Einstein was wrong on an even more fundamental basis.

"But when one starts with the wrong premise, no amount of patching can right the problem." (Seifer, 1996, 21)

The purpose of the Michelson and Morley experiment (1887) was to measure the movement of the earth through the ether. They didn't set out to prove or disprove whether or not the ether existed. That is why the results of their experiment greatly surprised the scientific community. Many theories and formulas were postulated in an attempt to explain the experimental results of the Michelson and Morley experiment. In 1905, Einstein's special relativity theory postulated that light didn't need a medium (ether) in order to travel through space. This created a dilemma. According the James Clerk Maxwell,

21

the speed of light was constant. If there was no ether, then the speed of light was constant to what? Einstein theorized that the speed of light was constant for all moving bodies, (all inertial frames). He based this assumption on Galileo's observation that the laws of physics are the same for all inertial frames. By carefully crafting the abstract concept of time into the Lorentz transformation, Einstein showed how this was mathematically possible. Einstein once said the following in referring to a mistake made by Max Planck:

> The main thing is the content, not the mathematics. With mathematics, one can prove anything. (Brian, 1996, 78)

Instead of exposing Newton's mistake and the motion myth, Einstein perpetuated them. This is where standing on Newton's shoulders tainted his reasoning when it came to the motion of objects through space. He treated the motion of objects as a whole instead of from their individual parts (atoms) working together. He built off the faulty paradigm that masses just mysteriously move through space with no atomic/quantum accountability for changes in speed or direction. In the end, both relativity theories fail to answer a basic fundamental question of physics: *How do masses move through space in the first place?*

Einstein's real blunder was not seeing Newton's mistake.

[Special Relativity is addressed in Appendix A: The Problem with Special Relativity. General Relativity is addressed in Part III under the subheading: And how does emitted energy from spherical bodies better explain the warped space of general relativity?]

Quantum Physics Misguided
What is the Role of an atom in the Momentum, Relativity, and Gravity of Masses?

"I like to think that the moon is there even if I am not looking at it"
Albert Einstein

The profound influence of Newton's mistake may have had its greatest detrimental consequence at the inception of quantum physics. At the time when scientists were exploring the composition of an atom and its role as the building block of all matter, they should have also considered the role of atomic/quantum processes in the motion of masses through space. With the scientific belief that all masses originate on an atomic/quantum level, is it not logical to think that the motion of masses through space also originates on the atomic/quantum level?

The unsuccessful equations Einstein jotted down and scratched out with his dying hands testify that the archaic belief that masses just mysteriously move through space is grossly insufficient in a science enlightened by atomic/quantum processes. This faulty paradigm that mass just mysteriously moves through space is still the predominant paradigm that is knowingly or unknowingly weaved into all present theories.

Two of our greatest physicists, Newton and Einstein, viewed the motion of objects through space from a whole perspective and not from their individual parts, perpetuating the motion myth. Quantum physicists, standing on their shoulders, also failed to see the underlying role of atomic/quantum processes in the motion of masses through space. This erroneous paradigm that masses just mysteriously move through space has now been passed down to us and continues to stifle our ability to see the underlying role of atomic/quantum processes in the momentum, relativity, and gravity of masses.

Quantum physics was misguided from its inception when quantum physicists naively ignored the underlying role of atomic/quantum processes in the motion of masses through space. Scientists are forward looking, waiting for the next breakthrough to add enlightenment to present knowledge and theories. Yet, general relativity and atomic/quantum processes will never be unified unless we look backwards to Newton's mistake and the motion myth and rebuild from there. The theme for part one of the book has been the following quote:

> But when one starts with a wrong premise, no amount of patching can right the problem. (Seifer 1996, 21)

If we want to understand momentum, relativity, and gravity from an atomic/quantum perspective, we must come to the conclusion that all objects are always in motion through space and that the individual atoms making up those objects are the source and cause of that motion. Then we will begin to clearly see from an inside-out perspective how atomic/quantum processes majestically orchestrate the perpetual motion of all masses through space. This is the grand key for unifying our present understanding of gravity (GR) with atomic/quantum processes.

At some point, we will all wonder why it took so long for the scientific community to surmise the absolute need for an atomic/quantum model of motion in order to explain how gravity really works.

Part II
—The Atomic Model of Motion—
The Role of the Atom in Momentum, Relativity, and Gravity

The Illusion of Rest

When Isaac Newton declared that an object in motion tends to stay in motion, while an object at rest tends to stay at rest, he probably didn't realize he was creating a false premise that would taint the minds of physicists for centuries to come. The "illusion of rest" is the greatest stumbling block to a viable explanation of how relativity and gravity operate from an atomic/quantum perspective.

The Atom and Motion: (*Each atom is a self-regulating confinement of energy that self-regulates changes in momentum though energy adjustments.*)

How does an object in motion stay in motion? We know that a car in motion eventually runs out of gas and stops. Why doesn't an asteroid orbiting the sun run out of gas? It just keeps going and going and going. What fuels the inertia of masses? In other words, how does an object in motion stay in motion? Do masses, such as an asteroid orbiting the sun, just mysteriously stay in motion through space without any explanation other than external forces can speed it up or slow it down? Masses get their motion from the very atoms that form their structures.

Atoms don't just mysteriously move through space. The energy making up an atom perpetually fuels its momentum through space, even when it appears to be at rest in a mass on the earth's surface. Any change in an atom's momentum is accompanied by an energy change to compensate exactly for the momentum change.

This single point, that **each atom is a self-regulating confinement of energy that self-regulates changes in momentum though energy adjustments,** forces us to rethink the role of the atom in the momentum, relativity, and gravity of masses.
The Atomic Model of Motion

The Atomic Model of Motion explains the role of the atom in the momentum, relativity, and gravity of masses. The atomic model of motion also sets the foundation for understanding how energy is recycled through spherical bodies.

The Atomic Model of Motion Overview

1. <u>Quantum Momentum</u> (or the role of the atom in the momentum of mass) is the equilibrium state of momentum where energy is neither absorbed nor emitted. This is the reason that an object in motion stays in motion.

2. <u>Quantum Adjustments</u> (or energy adjustments) are changes in momentum (acceleration and deceleration) that are accompanied by the absorption or emission of energy to exactly compensate for those changes. The science of motion is as exact as the science of chemistry. All changes in momentum can be explained by the addition or subtraction of energy.

3. <u>Quantum Relativity</u> (or how the energy of atoms and masses change as they go from one inertial frame to another) explains Galilean relativity from an atomic perspective. The sustaining energy driving all inertial frames is measurable and accountable on an atomic level using an understanding of quantum momentum and quantum adjustments.

4. <u>Quantum Gravity</u> (or the role of the atom in gravity): The acceleration of mass we call gravity is nothing more than the absorption of energy into the atoms of that mass. This accelerates already moving mass (atoms) in the direction of absorption. An acceleration of atoms is an acceleration of mass.

---

1. Quantum Momentum: The Role of the Atom in the Momentum of Mass

[*Quantum momentum is the reason why an object in motion stays in motion. The momentum of each individual atom drives the momentum of any given mass.*]

The atom is the never-ending energy that keeps all mass in never-ending motion. This is how planets continuously orbit the sun without having to be rewound or refueled. Even objects appearing to take a break on earth's

26

surface still cruise at nineteen miles per second as gravity yields them earthbound. Turn off gravity and these objects are no longer attached to the earth's surface. Give them a little nudge and they will float off in uniform bliss until an interaction with some other form of energy redirects their route or changes their speed.

Quantum Momentum is the equilibrium state or uniform motion of any atom and is the foundation for all motion from the simplest atom to the greatest of masses. The motion and momentum of any given atom is independent of the motion and momentum of all other atoms. It is the bonded relationship of the independent motion and momentum of the atoms making up a mass that gives that mass its motion and momentum through space.

From this point forward, I will periodically use the theme of yoked horses to explain the role of the atom in momentum, relativity, and gravity.

The best way to describe the **Quantum Momentum** of an atom is to take an adventure into the nucleus of an atom. Visualize a proton and a neutron yoked together like two horses yoked together pulling a carriage. Visualize the movement of the atom as the movement of these two horses. The yoke acts like the strong force that synchronizes their independent movements. With the snap of a whip or a pull on the reigns, the horses speed up, slow down, or change directions. Now imagine that these horses never stop moving. When forces act upon them, they speed up, slow down, and or change direction, but like the horses pulling Zeus' chariot, they never run out of energy.

In the emergence of an atom, energies form the atom's structure (yoking the horses together) while simultaneously engaging the atom's momentum through space, (the synchronized movement of the two horses when yoked). The two, structure and momentum, are inseparable.

As atoms form bonds with other atoms to create masses, the motion of each atom is linked together with the motion of other atoms to create the motion of the mass. The atoms that make up any mass share an order of movement that orchestrates that mass's movement through space.

A good way to visualize the **quantum momentum** of mass is to think about objects in momentum in the space station while it is orbiting the earth. Because the space station is in a free-fall, the independent momentum of all objects can clearly be observed. When an astronaut pushes an object, such as a bag of water, he is pushing all the atoms that make up that bag of water. When the force ceases, the atoms settle into the equilibrium of uniform motion. The atoms of the bag of water continue in the same direction until they bump into a wall of the station. As they run into the wall, a chain reaction transpires (which will be discussed in detail in the next section) until each atom experiences a momentum shift that leads to a new speed and direction. This new path continues until a force again acts upon all the atoms making up the bag of water. These atoms continue their new momentum uninterrupted even if the space station were to suddenly vanish from around them.

Any mass that appears to be at rest is just sharing the same motion with the object it appears to be resting on. Objects that the astronauts place in the air next to them—such as a toothbrush—stay in the same place. This is because the atoms making up the toothbrush have the same momentum through space as the atoms making up the astronaut. Their momentums are synchronized, moving at the same speed and in the same direction. Another example of the shared uniform motion of separate masses not connected to each other is objects within a car but not connected to it. As the vehicle accelerates, all the objects within the car go through internal changes to accommodate for the acceleration of the vehicle they appear to be resting in. At each new speed, all the objects in the car obtain the same uniform or quantum momentum as the car. When the car crashes into something, as its momentum is instantly changed, each object that appeared to be resting in the car continues in their own independent momentum until they crash into something such as the back of the seat, the dashboard, or the windshield. Each object has its own unique inertial momentum independent of all the other objects, including the car itself. Hence the importance of seat belts that physically attaches our momentum with the momentum of the vehicle in which we are riding.

What makes the quantum momentum of atoms and masses difficult to grasp is we continually see objects at rest on the earth's surface with no apparent motion whatsoever. We see rocks that have settled to the ground, furniture resting in houses, or a computer sitting on a desk. All of these seem to have no motion or momentum. The computer I am typing on does not appear to have any momentum so how could the atoms that make it up have momentum? This appearance of objects at rest is the same illusion that fooled Galileo and Newton and is still fooling physicists to this present day. The computer I am typing on is in a continuous state of acceleration towards the earth and only appears motionless because it is temporarily experiencing something similar to terminal velocity. Due to the effects of gravity, the uniform motion of the computer's atoms is riding on the uniform motion of the atoms of the earth. If the effects of gravity could temporarily be shut off, the momentum of the atoms of the computer through space would be more apparent. The atoms of the computer would continue to move through space at the same momentum of the atoms of the earth without being connected to the atoms of the earth because they would no longer be accelerating into them. Now if you gave the computer a push, it would slowly drift away from the earth, accentuating its own momentum through space, the synchronized momentum of the atoms from which it is made.

Quantum momentum is the inertial momentum of mass through space, which is caused by the synchronized momentum of the atoms from which mass is made. The speed and direction of mass through space remains unchanged just as Newton observed until an unbalanced force or energy acts upon the mass. Then the speed and direction are altered, but never at any time do the atoms of that mass stop or rest.

So what is the difference between an object at rest and an object in motion? Nothing. A body at rest shares the same motion as the body it appears to be resting on but their momentums through space are independent of each other. Just shut off the effects of gravity to prove this point.

2. Quantum Adjustments: Think of this section as a brief intermission that will be a useful tool to better understand quantum relativity and quantum gravity.

Quantum Adjustments are about the uneven forces that accelerate or decelerate mass. When mass accelerates or decelerates, quantum adjustments transpire within each atom to compensate exactly for these changes.

The momentum of mass does not change—a brilliant observation by Newton— unless acted upon by a force. The result of that force is a change in momentum. As changes in momentum occur, energy adjustments also transpire. Like the science of chemistry, the science of motion is exact. We just haven't figured it out yet like we have for chemistry.

As an atom's momentum increases, the atom will adjust its energy to exactly correlate with that atom's new momentum, and if an atom's momentum decreases, the atom will adjust its energy to exactly correlate with that atom's new momentum. These adjustments reestablish equilibrium within the atom as it goes through momentum changes. When the atom is in a state of equilibrium, that atom is in uniform motion and will maintain that motion until acted upon by another force or energy.

During momentum changes, each atom independently absorbs or emits the exact amount of energy to compensate for the momentum change. This regulates the momentum of an atom through space, whether the atom is by itself or is part of a collection of atoms in the form of a mass. All atoms experience quantum adjustments in energy in direct response to changes in momentum. This correlates perfectly with the conservation of mass-energy.

Quantum adjustments can better be explained by viewing each atom with a corresponding momentum pattern and energy level. The momentum pattern and energy level of an atom are two sides of the same coin that help explain changes to the continuous momentum of atoms through space. During the formation of an atom, protons, neutrons, and electrons unite in a synchronized dance to form and maintain the structure and motion of that atom through space. This structure, the atom's *momentum pattern*, describes the atom's perpetual momentum through space. The *momentum pattern* of an atom remains constant unless unbalanced forces act upon it. For this reason, I refer to an atom's uniform motion through space as a *momentum pattern* because this pattern continues until it is disrupted. As an atom's momentum changes, energy is proportionately absorbed or emitted (adjusted) to compensate exactly for the momentum change. Quantum adjustments are necessary to exactly compensate for disruptions to an atom's *momentum pattern*.

A good way to describe a *momentum pattern* is to go back to the visualization of horses yoked together. Think of one proton and one neutron as a pair of horses yoked together side-by-side walking at the same speed and in the same direction. Their combined movement is their momentum pattern. If we added additional pairs of horses to the first pair, so that each pair was lined up like a team of dogs pulling a sled, the combined movement of each pair of horses would be the momentum pattern of the combined team. Notice that no matter how many sets of horses we yoke together, each individual set of horses move at the same speed and in the same direction as all the other sets of horses, but with each additional set of horses added, the overall pulling power increases. Thus, the momentum pattern describes the speed and direction of the yoked horses, notwithstanding the number of sets.

Like the yoked horses, the nucleus of an atom could be described as yoked energies that simultaneously make up an atom's structure (the yoked horses) and momentum (the speed at which they are moving) through space. This energy, the energy that makes up the protons and neutrons of the nucleus of an atom, accounts for the motion of that atom through space. Like electrons, which can absorb and emit energy, protons and neutrons absorb and emit energy that regulates their momentum through space. As the linear speed of an atom increases, energy is simultaneously absorbed into every proton and neutron of that atom. As the linear speed of an atom decreases, energy is simultaneously emitted from every proton and neutron of that atom. (The electrons play a vital role in keeping the balance of energy within an atom by absorbing and emitting energy into and out of the nucleus and also into and out of the atom.)

The speed of any atom through space is determined by the *energy level* of each proton and neutron that makes up its structure or *momentum pattern*. As a force increases the speed of an atom, energy is absorbed into each proton and neutron. This increases the *energy level* of each proton and neutron of that atom. The same atom will now have an overall higher *energy level* per proton and neutron. Conversely, as a force decreases the speed of an atom, energy is emitted from each proton and neutron. This decreases the energy level of each proton and neutron. The same atom will now have an overall lower *energy level* per proton and neutron. The *energy level* of an atom (the speed the horses are moving) is in direct proportion to the energy level of each proton and neutron of an atom.

Energy level could be used in two different ways. There is the overall energy level of an atom (the total horse power) and the *energy level* of each pair of a proton and a neutron of an atom (the actual speed each individual horse is moving, which is also the same for all the horses). There is an important distinction. Let us compare an atom with one proton and neutron pair (one team of horses) to an atom with five pairs of protons and neutrons (or five teams of horses). If these two different atomic size atoms have the same momentum through space, then the *energy level* of each pair of protons and neutrons is equal (all the horses are moving at the same speed), but the overall energy level or amount of energy (total horse power) will be greater for the atom with five pairs than the atom with one pair. It will have about five times the overall energy, even though the atoms move through space at the same speed. The atom with the five pairs will have more inertia or resistance to change than the atom with one pair. Throughout the rest of this book, when I refer to the *energy level* of an atom, I am referring to the energy level of one pair of a proton and neutron. This is the *energy level* that describes its linear speed through space, notwithstanding its atomic number.

The protons and neutrons ability to absorb and emit energy allows for an atom to adjust its momentum when a force acts upon it. This holds true for each atom, notwithstanding the number of protons and neutrons in its nucleus. As the atom increases in speed or changes direction, energy is absorbed equally into every proton and neutron of the nucleus of that atom (all the horses start running faster). As the atom decreases in speed, energy is emitted equally from every proton and neutron of the nucleus of that atom (all the horses slow down). This is how the energy level (speed of one horse) and momentum pattern (the equilibrium state or steady speed of the entire team) are inseparably connected. A change in the momentum pattern of an atom (a force speeds it up or slows it down) initiates a simultaneous change in the atom's energy level (the speed at which each horse moves).

The horse example shows how the momentum pattern and energy level are actually a single process. As the *energy level* of each horse increases, the *momentum pattern* of the entire team increases by the same amount, and vice-versa. Imagine you have a team of six horses, with two horses yoked side by side, with another set yoked behind them, and the last set yoked behind them. The speed that each horse moves is the *momentum pattern* for that whole team. They all move at the same speed. When you apply a force, such as a whip to speed them up, or a pull on the reins to slow them down, you simultaneously change the energy level (speed) of each horse and the momentum pattern (speed) of the whole team. The *energy level* of one horse is the same as the *momentum pattern* for the entire team. The energy level and momentum pattern for one set of a proton and a neutron is the same energy level and momentum pattern for all the other sets of a proton and a neutron within the same atom. When you change the momentum pattern, the energy level simultaneously changes. And vice versa, when you change the energy level, the momentum pattern simultaneously changes. They are two sides to the same coin.

Mass is a collection of synchronized atoms moving through space as the momentum pattern of each atom moves in concert with all the other atoms of that mass. Mass maintains a constant speed and direction until there is a disruption to the momentum pattern of each atom that makes up that mass. When the momentum patterns of the atoms of a mass are disrupted, energy is absorbed or emitted from the mass, from each individual proton and neutron of each atom. This reestablishes equilibrium within each atom within the mass to accommodate for the new speed and direction of that mass through space.

Visualize two masses colliding. The *momentum patterns* of the atoms of each mass are disrupted. As the *momentum patterns* are altered, the accompanying *energy levels* of the protons and neutrons of the atoms of each mass simultaneously adjust to accommodate the new *momentum patterns*. An absorption or emission of energy to exactly coincide with the degree that the *momentum patterns* were altered transpires, (each horse slows down or speeds up). As equilibrium is restored, the masses continue in their new momentums until a force or energy acts upon them again.

In summary, the momentum pattern of an atom and energy level of each proton and neutron that makes up the nucleus of an atom regulates the speed and direction of that atom through space. When momentum changes occur, quantum adjustments transpire to reestablish the equilibrium of quantum momentum. Since mass is the synchronized movement of individual bonded atoms, a momentum change in mass is a momentum change for every atom that makes up that mass. Every atom experiences a quantum adjustment. As there is no *something from nothing* in the cause and effect science of physics, changes in the momentum of mass must be accounted for by corresponding energy changes within that same mass.

3. Quantum Relativity: How Atoms and Masses Change as They Go from One Inertial Frame to Another

Galileo learned that if a person were in a closed cabin of a ship moving at a uniform speed, he wouldn't be able to tell his relative motion to the land by an appealing to the laws of physics. If he dropped an object, it would fall straight down. If he threw it up in the air, it would go straight up and straight down. In all, there was nothing he could do to determine his relative speed to land by applying what he knew about the laws of physics.

On the surface, the laws of physics appear to be the same for all inertial frames, but a deeper look into the microphysics of relativity will reveal differences on an atomic/quantum level for the same mass in different inertial frames. Since Einstein liked to use trains to explain special relativity, I will use a train example to explain quantum relativity.

If person A, traveling on a train holds a beanbag four feet above the floor and drops it to the floor of the cabin wherein he is riding, he will measure the beanbag to have fallen four feet. If person A decides to drop the same beanbag out of the moving train's window four feet above the ground, he will

34

see the beanbag fall four feet straight to the ground in the same manner that he saw the beanbag fall four feet straight to the floor in the cabin of the train. Person B, standing on the stationary earth relative to the moving train, will measure the distance of the falling beanbag to fall more than the four feet as observed by person A. For person B, the beanbag will not only fall four feet to the ground as measured by person A, it will also fall at an angle proportional to the speed of the train. For this reason, each observer will literally measure the beanbag falling different differences. (See illustration—1). This is classic Galilean relativity, but it brings up an interesting dilemma: How can the same falling object literally travel multiple different distances such as this falling beanbag dropped by person A and observed by person B? The answer to this question will be found in the *momentum pattern* and *energy level* of each atom of person A, person B, and the beanbag.

Illustration--1

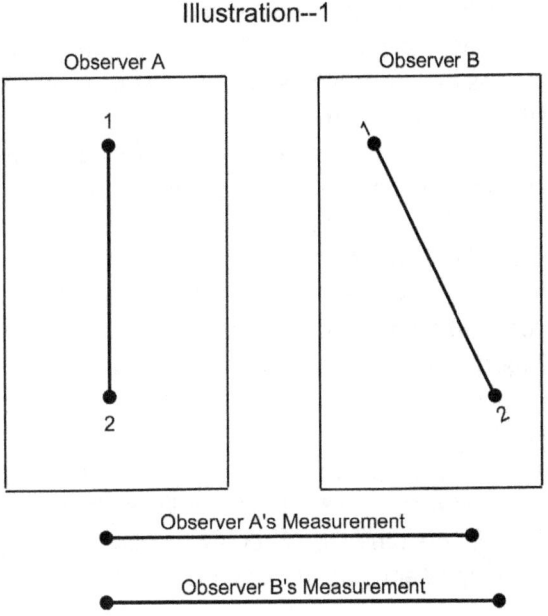

With each incremental increase of speed to the beanbag as it travels in a moving train—relative to earth's inertial frame—the more overall energy the beanbag acquires. (Each horse runs a little faster—energy is absorbed into each atom making up the beanbag.) For example, if person A drops a beanbag four feet into a pile of sand outside of a non-moving train, the beanbag will displace a certain amount of sand upon impact. The energy acquired by the falling beanbag due to the effects of gravity is transferred to the sand at impact. Now, if the train is moving and person A drops the beanbag the same four feet so it hits the same sandbox that is outside of the train, falling at an angle due to the movement of the train, the overall amount of displacement of sand caused by the falling beanbag is increased. The beanbag literally contained more energy to be transferred to the sand on impact, creating a greater displacement of sand.

This increase of energy is an actual increase of energy that is added within the very structure of every atom making up the beanbag. Any change in momentum is a change in the overall energy level of a given mass. This means that the same beanbag in different inertial frames has differing amounts of mass-energy.

How can two different observers, such as observer A and observer B, measure the same falling object to travel two different distances? The difference in distances can be explained on the atomic level. The atoms of the beanbag dropped by person A had the same momentum pattern as the atoms of the moving train and person A on the moving train that dropped the beanbag. The atoms of person B, who was standing on the earth watching the train go by, had the same momentum pattern as the atoms of the earth upon which person B was standing. The atoms of the beanbag, person A, and the train were at a higher energy level than the atoms of person B and the earth. Person A observed the beanbag from the same energy level as the beanbag, while person B observed the beanbag from a lower energy level than the beanbag. Being at the same energy level as the beanbag, person A sees the beanbag fall straight down as observed in Galilean relativity. Being at a lower energy level than the beanbag, person B sees the beanbag fall at an angle towards the ground. From their immediate energy levels in relation to the energy level of the beanbag, the measurements of both observers is absolutely correct even though the beanbag took only one absolute path through space.

When the beanbag was dropped from the moving train, the atoms of the beanbag maintained the momentum pattern of the moving train until impact. Upon impact, each atom of the beanbag experienced a quantum adjustment as they switched inertial frames. In this case, the momentum pattern of each atom of the beanbag lost energy as they adjusted to become synchronized to the new momentum pattern of the atoms of the earth.

In summary, the faster a beanbag (or any mass) moves relative to the earth's inertial frame, the more energy it acquires. The energy level of mass changes when the inertial frame that carries it changes. As a train (or any mode of transportation carrying mass) accelerates to go from one inertial frame to the next, say from 5 mph to 10 mph, its mass (and any mass that appears to be resting on it) literally acquires more energy within its atomic structures. This causes observers of differing energy levels to observe the motion of the same mass differently.

In Stephen Hawking's book, *A Brief History of Time*, Hawking states, "…suppose our Ping-Pong ball on the train bounces straight up and down, hitting the table twice on the same spot one second apart. To someone on the track, the two bounces would seem to take place about forty meters apart, because the train would have traveled that far down the track between the bounces….The positions of events and the distances between them would be different for a person on the train and one on the track and there would be no reason to prefer one person's positions to the other's" (Hawking, 1988, 17-18.) What Hawking is admitting here is that there is no scientific explanation accounting for the differences in the distances as observed by both observers. Quantum relativity offers an explanation. Each person from their own inertial frame will observe the path of the Ping-Pong ball through space differently because the mass/atoms of each inertial frame has a different momentum pattern and energy level. The person in the train shares the same momentum pattern and energy level as the train and the Ping-Pong ball while the person on the track observes from a different momentum pattern and energy level than the train, the person on the train, and the Ping-Pong ball. Again, the cause of this phenomenon is that observers of differing energy levels will observe the motion of the same mass differently.

Why do objects that can appear to be at rest within an inertial frame follow the same laws of physics? Whether the train is at rest at the station or traveling at a uniform speed of 100 miles per hour, the Ping-Pong ball falls straight to the ground. What's going on inside the Ping-Pong ball?

Let's put an *atom* on the Ping-Pong table. Because the *atom* seems to be resting on the Ping-Pong table, one might assume that the imaginary horses within our *atom* are at rest, too, that when the train picks up speed, the horses can stand there and enjoy the ride. In other words, when the train goes from zero to 100, one might assume that nothing transpires within the *atom*; it is just a passive passenger along for the ride. But from an atomic model of motion perspective, when our imaginary horses within the *atom* were on the Ping-Pong table at the station, their legs were moving. The momentum pattern of the *atom* is the same as the momentum pattern of the atoms making up the Ping-Pong table, and the Ping-Pong table has the same momentum pattern as the train stationed on the tracks, and the tracks have the same momentum pattern as the earth, to which they are attached. As a force increases the speed of the train, all the atoms that are a part of the train experience changes to their momentum patterns and energy levels. Even the *atom* that appears to be resting on the Ping-Pong table experiences changes in its energy level and momentum pattern as the speed of the train increases. The imaginary horses of the *atom* resting on the Ping-Pong table change their energy level and momentum pattern to match the energy level and momentum pattern of each pair of yoked horses of all the other teams (atoms) connected to the train. All the horses of all the atoms increase their energy level. (This is what is meant by there are no free rides.)

In other words, the Ping-Pong ball is not a passive passenger on the train. As the atoms within the train increase in energy, the atoms of the Ping-Pong ball will also increase in energy. Because the momentum pattern of the atoms of the Ping-Pong ball remain synchronized with the momentum pattern of the atoms of the Ping-Pong table and the rest of the train, the Ping-Pong ball will drop straight down if you held it above the table and dropped it. The effects of gravity cause the ball to fall straight down. This phenomenon happens within any inertial frame where all the atoms share the same momentum pattern (the horses are all moving at the same speed and in the same direction).

To help visualize relativity from an atomic model of motion perspective, imagine the following experiment that could be done on the international space station that is free falling around the earth. Imagine Person A floating from one side of the space station to the other side. When he reaches person B, who is stationed in the middle, he lets go of a quarter. Because the individual atoms that make up person A and the quarter share the same momentum pattern and energy level (even after person A lets go of the quarter) person A and the quarter continue to move through space at the same speed and in the same direction. Person B, who is stationed in the middle, watches as the quarter slowly moves away at the same speed that person A is moving away. Person B sees the quarter travel a greater distance from his immediate perspective than person A, who sees the quarter travel with him as if he were still holding it. Even though the quarter travels only one distinct path through space, each observer measures a different length of travel of the quarter from his or her

38

immediate perspective. This is because Person A and the quarter remained at the same energy level, while person B observed from a different energy level.

If Galileo could have measured the energy level of the atoms of any object in his boat cabin while his boat was tied to the dock, he then could have compared that measurement to the energy level of the atoms of the same object when his boat was sailing. The measured difference would validate that as mass goes from one inertial frame to another, the momentum pattern and energy level of the atoms of that mass change to accommodate the new uniform motion. The acceleration and deceleration of masses are only temporary stages between the equilibrium of uniform motion. The acceleration or deceleration of mass is always accompanied by the absorption or emission of energy. This allows each atom to adjust to the new energy level required to maintain its new momentum pattern.

Quantum relativity is the reason the laws of physics are the same for all inertial frames.

4. Quantum Gravity: The Role of the Atom in Gravity

Gravity is typically explained as an attraction between masses. Quantum gravity explains this attraction as the emission and absorption of energy between masses. The absorbed energy accelerates the momentum of atoms. An acceleration of atoms is an acceleration of mass.

Is an object that is sitting on the earth's surface with no apparent motion really at rest? Newton's observation of distinguishing between objects at rest verses objects in motion has proven fatal for many centuries in understanding how gravity really works. As a result of Newton's mistake, the motion myth paradigm (the myth that mass just mysteriously moves through space) has tightly gripped the minds of most, if not all, physicists. And, to this day, physicists cannot explain how gravity operates from an atomic/quantum perspective.

When Newton said that an object in motion tends to stay in motion unless acted upon by a force, he was obviously referring to an external force such as a collision or friction. The result of the external force is a change in the momentum of that mass. From an atomic/quantum perspective, an external force causes a change in the momentum pattern of each atom, which initiates a simultaneous change in the energy level of that atom. This happens every time an external force acts upon an object in motion. Energy is absorbed or emitted to exactly correlate with the momentum change.

The same effect occurs when energy is absorbed into the nucleus of an atom. A change in the energy level of an atom initiates a change in the momentum pattern of that atom. [Say the horses are spooked and they start running faster. This is a change in the *energy level* of each horse. In the same instance, the overall speed of all the yoked horses increases. This is a change in the overall *momentum pattern* of that team of horses.] As energy is absorbed into each proton and neutron of an atom, the result is the acceleration of that atom in the direction of absorption. Its momentum simultaneously shifts to accommodate the increase in energy. This momentum shift is the acceleration.

Einstein's happiest thought was the realization that inertial acceleration and gravitational acceleration are equivalent. In other words, gravity is the acceleration of mass. He used an example of an elevator being pulled up in empty space at 32 feet per second per second to illustrate this point. (Thirty-two feet per second per second means that for every second of acceleration, an object travels an additional 32 feet of the distance of the previous second. At the end of the first second, an object travels 32 feet. For the second second, the object travels 64 feet within that second for a total of 96 feet for the first two seconds (32+64). For the third second, the object travels 96 feet within that second for a total of 192 feet (96+96) for the first three seconds and so on.) The elevator being pulled up in empty space would create the same gravitational acceleration that a person experiences standing on the earth or an object falling towards the earth. The reality that gravity is the acceleration of mass creates the basis for quantum gravity.

From an atomic perspective, inertial acceleration is equivalent to gravitational acceleration. Both result in the change of the energy level and momentum pattern for every atom involved. An external force causes inertial acceleration, whereas the absorption of energy causes gravitational acceleration. Whether by an external force or the absorption of energy, the result is the same—the atom or mass accelerates. This is because a change in the momentum pattern (external force) changes the energy level of an atom, and vise versa, a change in the energy level (absorption of energy) changes the momentum pattern of an atom. Either way, the result is acceleration. The simultaneous effect of the momentum pattern changing the energy level or the energy level changing the momentum pattern is like blowing air into a balloon. As the balloon receives more air, the boundaries of the balloon expand. As the balloon loses air, its boundaries contract. As pointed out in the quantum adjustment discussion, an external force can accelerate or decelerate the linear speed of an atom. Energy absorption always accelerates the linear speed of an atom.

Since mass is a collection of bonded atoms, an acceleration of atoms is an acceleration of mass. The constant flow of energy into the atoms of a mass creates gravitational acceleration. The absorbed energy changes the energy level of each atom, which simultaneously changes the momentum pattern of each atom. This accelerates the already moving mass in the direction of absorption.

This is why Newton's misperception was so crucial. If mass is perceived to be at rest when it is sitting on the earth's surface, then the atoms making up that mass are not perceived to be the cause of that mass's motion through space. Newton's misperception assumed that mass is getting a free ride on the earth. This misperception blocks the potential for understanding how gravity really works. By correcting Newton's mistake, we now understand that the atoms that make up any mass are the cause and continuation of that mass's motion through space, (the horses are always moving). What appears to be mass resting on the earth's surface is actually mass in motion wanting to accelerate but experiencing terminal velocity due to the resistance of the earth's surface. If we could shut off the effects of gravity—the energy causing acceleration—the mass would no longer be accelerating into the earth, nor would it be attached to the earth. A good nudge would cause the mass to go off in a different direction than the earth, accentuating the motion it already had.

The effects of absorbed energy causing gravitational acceleration can be seen when an accelerating object decelerates at the moment of impact with a lower energy level surface such as the earth's surface. In the process of reestablishing equilibrium with atoms of a lower energy level, quantum processes transpire. This sets off a chain of events that can be witnessed at the moment of impact as will be explained in the falling penny example.

The Falling Penny

An example of the effects of quantum gravity is dropping a penny to the floor. First of all, as you hold the penny above the floor, the penny is already in motion. It is sharing the same motion as the hand holding it, just as the quarter shared the same motion as the moving astronaut, even after he let it go. When you let go of the penny, the atoms making up the penny absorb energy emanating from the surface of the earth. (The energy emanating from the earth's surface will be addressed later.) The energy level of each atom changes—like the snap of a whip causing each horse to run faster. This simultaneously changes the momentum pattern of each atom, (the whole team is moving at a faster pace). This increases the overall motion of the penny in the direction of absorption 32 feet per second per second until it hits the ground.

When the penny hits the ground, its increased energy level from the absorbed energy that accelerated its momentum interacts with the lower energy level of the ground. At the very moment of impact, the momentum pattern of each atom of the penny changes. Energy is emitted from each atom to accommodate each atom's new momentum pattern. At the initial bounce, the penny is off in a different direction. If not for gravity, the penny would appear to float off in its new direction until a force acted upon it. (To help visualize this, think of an object in the space station that is floating horizontally towards one of its walls. When the object hits the wall, it changes its direction, and then continues in its new direction until a force acts upon it.) As the penny starts to go off in a different direction, it begins to absorb more energy emanating from the earth's surface, changing its direction and accelerating it towards the earth again to repeat the process over and over until the penny finally appears to rest on the earth's surface, with its atoms sharing the same momentum pattern as the atoms making up the earth. During this whole process, the sum of mass-energy in the universe remains unchanged.

When the penny makes contact with the lower energy level surface, there is an immediate change in the energy level of every atom of the penny. The atoms of the surface where the penny landed are also disrupted at impact, causing the absorption and emission of energy according to each atom's equilibrium need. (This is like the sand that was displaced in the sandbox when

42

the beanbag landed in it.) When the penny finally appears to rest on the earth's surface, the momentum pattern and energy level of the atoms of the penny remain synchronized with the momentum pattern and energy level of the atoms of the earth until another disruption occurs such as when somebody picks up the penny.

Why does the penny appear to stay at rest on the earth's surface? Even though it is continuously absorbing energy emanating from the earth's surface, causing it to want to accelerate, its continuous contact with the surface of the earth restricts the acceleration. In other words, it has reached terminal velocity. The energy necessary for acceleration is continuously being absorbed into the penny, but because the momentum pattern of the atoms cannot change due to their terminal velocity, the absorbed energy is emitted at the same rate of absorption.

Drop a penny a few times on the surface of a table. Watch at impact as the penny goes from a higher energy level (the energy acquired to accelerate it towards the surface of the table) to the lower energy level of the surface of the table (the previous energy level of the penny before you picked it up and dropped it). Drop the penny from various heights to compare the amount of energy it acquires from the varying heights. As the penny hits the table and bounces a few times and then vibrates until it eventually comes to rest on the table, you are witnessing the atoms of the penny going from a higher momentum pattern and energy level to the lower momentum pattern and energy level of the atoms of the table.

Gravity is not a mysterious force of attraction between masses, nor is it the effects of warped space-time, but rather, it is energy (forces) acting within matter. It is the effect of absorbed energy on already moving mass. This is why exposing Newton's misperception that objects can be at rest is so important to understanding how gravity really works. Without this step, you cannot understand how gravity is the absorption of energy accelerating mass that already has momentum.

This is why Einstein was unable to unify general relativity (his explanation of gravity) with atomic/quantum processes. Like Newton, he failed to recognize that all objects are always in motion, the first major step for understanding how gravity operates from an atomic/quantum perspective. Instead, he brilliantly created space-time as a substitute for atomic/quantum processes, accurately predicting the effects of gravity without addressing the atomic/quantum causes.

Space-time describes distortions in space caused by large masses, such as suns and planets, and the effect this has on moving objects near their surfaces, such as other planets, moons, asteroids, and light. In reality, these distortions are actually higher concentrations of energy emanating from the surfaces of large spherical masses such as the sun or the earth. (This will be explained in more detail later.)

Quantum gravity can now be easily explained as the acceleration of mass caused by the absorption of energy. The absorbed energy changes the energy level of each atom, which simultaneously changes the momentum pattern of each atom, causing acceleration towards the source of the energy being absorbed.

Because the momentum of each proton and neutron is independent of the momentum of all other protons and neutrons (like each set of yoked horses connected together with other yoked horses) the number of protons and neutrons in an atom doesn't change that atom's acceleration rate in comparison to atoms of differing numbers of protons and neutrons, (atomic number). All protons and neutrons of a mass are exposed to the same energy source and absorb the same amount of energy, causing the same momentum shift in the direction of absorption. (All the horses increase their speed at the same rate.) This is why Galileo could drop two differing size balls off the same tower and watch them land at the same time. Because the larger ball has more protons and neutrons, its overall energy or inertia will be greater than the ball with fewer protons and neutrons. But the acceleration of both balls will be the same because all protons and neutrons accelerate at the same rate, for all protons and neutrons are exposed to the same energy concentration emanating from the earth. For this reason, all atoms, elements, compounds, and masses accelerate at the same rate. This demystifies the equal attraction law of gravity.

The Visible Effects of Quantum Gravity

The absorbed energy due to the effects of gravity is visible. When I hold a pencil several feet above a hard tile floor and drop it, the absorbed energy responsible for accelerating the atoms of the pencil causes the pencil to bounce a few times when it hits the floor. Energy is emitted from the atoms of

44

the pencil with each bounce until the atoms of the pencil share the same energy level as the atoms of the surface of the floor upon which it now appears to be resting. When I held the pencil above the floor, it is said to have potential energy. Potential energy is nothing more than potential changes in momentum patterns—acceleration—before a terminal velocity is reestablished with the surface of the earth.

When a ball bounces off the ground, it moves away from the surface of the earth. It would continue in this direction uninterrupted if it weren't for energy being absorbed back into the atoms of the ball due to the effects of gravity. After the ball reaches its apex, briefly sharing the same momentum pattern and energy level as the atoms of the earth, it begins to accelerate towards the earth again. With each bounce, more energy is emitted until eventually the ball comes to a resting position on the earth. You can visually see the effects of acquired energy causing the acceleration that precedes each bounce. When the ball appears to be resting on the earth's surface, it is still moving through space. The momentum pattern and energy level of each atom of the ball are still the source and cause of its movement through space. If I shut off the energy emanating from the earth and then threw the ball into the air, it would move away from the earth like a helium balloon escaping from the grasp of a child.

As absorbed energy accelerates mass towards the surface of the earth or as emitted energy accompanies the deceleration of mass when it appears to come to an abrupt stop, the effects and results are observable and ultimately measurable.

The Imaginary Wall in Space

An example to help visualize quantum gravity is to imagine a wall in space emitting gravitational energy. If you placed a marble several feet away from the wall, it would begin accelerating towards the wall. When it reached the wall, it might bounce a few times and then the marble would appear to rest against the wall. The energy level and momentum pattern of each atom of the marble would be at a terminal velocity against the wall, wanting to accelerate but restricted by the wall. The atoms of the marble would share the same momentum through space as the atoms that make up the wall. In this state, the same applied force would move the marble the same distance in any direction on the wall. (In essence, the wall acts like the surface of the earth.) If you flicked the marble with the same force in any direction, it would go the same distance. If you picked it up and let it go, it would accelerate towards the wall. When it hit the wall, it would bounce a few times, emitting energy until the energy level and momentum pattern of the atoms of the marble matched the energy level and momentum pattern of the atoms of the wall. Then it would again be in a state of terminal velocity against the surface of the wall, appearing to be at rest on the wall.

Tides

Another effect of gravity that is observable but difficult to explain is the cause of tides. Heretofore explained as the gravitational pull of the moon on the earth can now be explained as concentrated gravitational energy emitted from the moon and absorbed by the atoms making up the water. This causes a shift in their momentum patterns, accelerating them in the direction from which the energy was absorbed until a terminal velocity is reached between the energy emitted by the moon in contrast to the energy emitted by the earth. The moon doesn't mysteriously attract the water nor does warped space cause the water to move towards the moon, but rather, it is energy emanating from the large, spherical-shaped moon that accelerates the atoms of the water towards the direction of absorption.

Satellites

Whether it is our sun orbiting the center of our galaxy, the earth orbiting our sun, or the moon or satellites orbiting the earth, quantum gravity keeps them in orbit. As the Galaxy Gravity Cycle (GGC) keeps the flow of energy moving through the galaxy, spherical bodies absorb this energy and then emit it through their surfaces. (This process will be discussed in more detail in next section of this book.) The energy emanating from their surfaces accelerates bodies already in motion (within a reasonable proximity of their surface) in an orbital pattern around their spherical shape. It is precisely because these satellites already have motion that energy emanating from the surfaces of large spherical bodies can accelerate their motion into an orbital pattern.

Summary

Gravity is the absorption of energy into the atoms making up mass, initiating and sustaining the acceleration of already moving mass in the direction from which the energy is being absorbed.

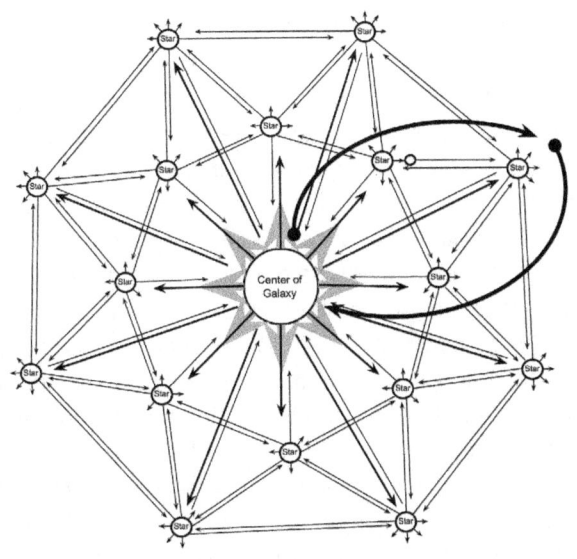

Part III
—The Galaxy Gravity Cycle (GGC)—
How Spherical Bodies Absorb and Redirect Energy within a Galaxy

[At the beginning of this book, a childlike explanation was given of the Galaxy Gravity Cycle (GGC). The following will go into greater detail concerning the role of spherical bodies in the Galaxy Gravity Cycle. The Galaxy Gravity Cycle (GGC) along with the Atomic Model of Motion (AMM) replaces the erroneous foundation of modern physics.]

How does the earth (or any spherical body) sustain the constant effects of gravity?

The earth, as a spherical body, is already in motion around the sun—quantum momentum. The amount of gravitational energy it receives from the sun allows it to continuously accelerate towards the sun without accelerating into it—quantum gravity. What happens to the gravitational energy absorbed into the earth? How does this energy create earth's gravitational effect? How does emitted energy from spherical bodies better explain the warped space of general relativity? And how does the Galaxy Gravity Cycle correct the erroneous foundation of modern physics?

What happens to the gravitational energy absorbed into the earth?

What happens to the gravitational energy that is absorbed into the earth that keeps the earth accelerating into the sun? The energy absorbed into the earth that keeps the earth accelerating into the sun is eventually emitted through the spherical surface of the earth, creating the effects of gravity that we experience living on the earth. How does this work? The core of the earth, or the core of any spherical body, is the catalyst that maintains the cycle of energy flowing into and out of spherical bodies. The atoms of a spherical core absorb and emit a continual flow of gravitational energy. As mentioned in quantum gravity, energy absorbed by an atom causes it to accelerate towards the direction the energy came from, (AMM). As atoms of the core absorb gravitational energy, they are restricted in the amount of acceleration they can experience. This is due to their proximity with the other atoms within the core. The energy not integrated into one of these confined atoms of the core (due to their inability to accelerate) is simultaneously emitted from that atom. The emitted energy can then be absorbed into a nearby atom. But as the nearby atom's acceleration is also limited, it simultaneously emits the energy not being integrated into it. This process repeats itself over and over again within the core of compactly confined atoms. The gravitational energy that has nowhere to be absorbed within the spherical core is emitted through the surface area of the spherical body and out into space. This gravitational energy emanating from the surface area creates the gravitational effects that spherical bodies exhibit near their surfaces. As the gravitational energy being emitted through the spherical surface goes out into space, it causes other masses to accelerate towards the spherical body that is emitting the gravitational energy. It is important to note that the concentration of gravitational energy will be greater near the surface of the spherical mass and will defuse as it goes further away from the spherical surface, (See Illustration—2 Energy Emitted from Spherical Surface Area). In Summary, as energy is absorbed into a spherical body, the atoms either accelerate or—like the penny experiencing a process similar to terminal velocity as it rests on the earth's surface—acceleration is restricted and energy is emitted at the same rate of absorption. The unabsorbed gravitational energy is emitted through the spherical surface out into space. This constant flow of emitted gravitational energy accelerates masses towards the spherical surface and keeps atoms and molecules earthbound.

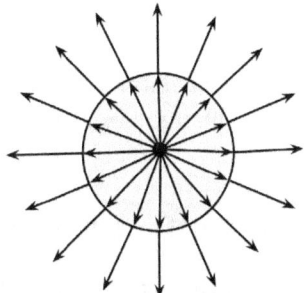
Why do the atoms near the surface of the earth stay earthbound? Shouldn't they just independently accelerate towards the sun? As the atoms of the earth facing the sun receive gravitational energy from the sun, they accelerate towards the sun. This is most evident with the fluidity of water in the case of tides. As water molecules receive gravitational energy from the sun and moon, they begin to accelerate towards the sun and moon. At the same time, the gravitational energy recycled through the core of the earth and emitted through the surface area of the earth accelerates the atoms on the surface of the earth towards the surface of the earth, keeping them earthbound. This simultaneous tug-a-war creates the phenomenon of tides and explains why tides are at their highest when the sun and moon are aligned but why the ocean waters (and all other atoms, molecules, and compounds) always stay earthbound.

How do spherical bodies increase in their size and mass? The gravitational energy emitted from a spherical core causes other masses to accelerate into the core. This builds up further mass around the core. As mass is covered by other mass, the atoms of the covered mass are restricted in their movement and simultaneously emit gravitational energy. This increases the circumference of the amount of atoms that are restricted in movement, increasing the spherical size and surface area emitting gravitational energy. This creates a greater concentration of energy (gravitational effect) near the growing surface area and a stronger gravitational field surrounding the spherical body. (The gravitational field is the concentration of gravitational energy being emitted from the spherical surface.) This process of the spherical core increasing in mass as mass accelerates into it describes how spherical bodies increase in size while simultaneously increasing their gravitational effect.

How does this energy create earth's gravitational effect?

The Cycle of Absorbed/Emitted Energy Creates a Spherical Body's Gravitational Effect

As stated above, the space necessary for atoms to change their momentum within a spherical core is limited. As a constant flow of energy accelerates atoms within a spherical core, the amount of acceleration they can experience is limited due to the compact nature of all the atoms making up the spherical core. Each atom of the core then experiences a process similar to terminal velocity. At terminal velocity, the atoms absorb and emit energy at the same rate. Eventually, the energy emitted by the limited accelerations of the atoms that make up the spherical core has nowhere to escape except through the surface of the spherical core, and when it reaches the surface, it is then emitted into space. This causes any object on or near the surface of the spherical surface to accelerate towards the surface. This cyclical process of spherical bodies (like the earth) absorbing and emitting gravitational energy keeps a constant flow of gravitational energy moving into and out of a spherical body. And the energy continuously being emitted through the spherical surface into space creates the gravitational effect associated with the spherical body. The denser the concentration of this energy being emitted, (due to a greater size and mass of the spherical body), the greater is the gravitational effect in accelerating atoms, molecules, and masses on or near the surface. And as the energy emitted through the surface of a spherical body continues to go further out into space, so does the gravitational effect of that energy. It will accelerate any atoms or masses into which it is absorbed. Again, Illustration—2 shows how the energy emitted from a spherical body is diluted as the energy spreads out the further it gets away from the spherical surface.

50

And how does emitted energy from spherical bodies better explain the warped space of general relativity?

The size and mass of a spherical body determines the amount of energy flowing into it. This in turn determines the concentration of energy exiting its spherical surface. The energy exiting a spherical surface is the warped space-time of general relativity. The gravity effect is greatest around the surface area of spherical bodies and weakens the further you move away from the surface. (See Illustration—2 Energy Emitted from Spherical Surface Area.) What Einstein mathematically composed as warped space-time in his general relativity theory is actually the effects of concentrated gravitational energy exiting through a spherical surface. As this energy moves away from the spherical surface, its concentration is diluted more and more the farther it moves away from the spherical surface, but when any of this energy is absorbed into the atoms of a mass, it will cause those atoms to accelerate. Einstein's general relativity theory is a mathematical description of higher and lower concentrations of gravitational energy exiting through spherical surfaces.

Electromagnetic Waves are not Impervious to Gravitational Energy

If, in the process of reading the previous paragraph, you are wondering about light waves bending around large spherical bodies, continue reading. This book claims that gravitational energy permeates space. This energy is absorbed and recycled by spherical masses (which are made up of atoms) just as electromagnetic waves are also absorbed and emitted by atoms. Gravitational energy, which is capable of being absorbed and emitted by the nucleus of an atom, is capable of being absorbed into electromagnetic waves moving freely through space. As a photon passes by a spherical body, it passes through a field of high concentrations of flowing gravitational energy being emitted through a spherical surface. Some of this energy is absorbed directly into the photon. Instead of increasing acceleration like it does when absorbed into the nucleus of an atom, the photon experiences a frequency shift. The photon now has more energy in the form of a higher frequency. The speed of the photon didn't change, but the frequency shift caused a slight altering of its path in the direction from which the energy was absorbed. Bent light around spherical bodies is the result of a frequency shift within photons.

Proof

The validity of the Atomic Model of Motion (AMM) and the Galaxy Gravity Cycle (GGC) hinges upon the starlight displaced by the gravity of the sun. Einstein pointed to the warping of space-time as the cause of this validated phenomenon, (Eddington, 1919). Quantum gravity, as explained in this book, theorizes that the bending of light around large massive objects like the sun is not caused by warped space, but rather, it is caused by large concentrations of gravitational energy emanating from the surfaces of spherical bodies like the sun. As light passes the surface of the sun, each photon absorbs some of this energy, causing a frequency shift. Without affecting its speed, this frequency shift slightly changes its direction towards the source that initiated the shift. The Atomic Model of Motion theorizes that the displaced starlight will be blue shifted from the same starlight that isn't displaced, validating the underlying principle from which the Atomic Model of Motion (AMM) is built upon: Energy transfers accompany changes in speed or direction, otherwise, the conservation of mass-energy would be violated.

How Does the Galaxy Gravity Cycle Correct the Erroneous Foundation of Modern Physics?

It is a hard pill to swallow to think that modern physics is built on an erroneous foundation. But if you think about it carefully, it really isn't that difficult to conceptualize. Classical physics enjoys a continual progression up to Einstein's general relativity. Just prior to general relativity, an understanding of atomic/quantum physics begins to emerge. In the emergence of atomic/quantum physics, there is a clear and distinct break from classical physics. The new discoveries in atomic/quantum physics didn't piece together well with the traditional, high-esteemed classical physics that had the signature of Isaac Newton. In this split, a new science emerges that is distinct and separate from the old science. Yet, the new science cannot account for all the actions and reactions that take place on a larger scale. Thus, in physics, we are left with a science for the macro, (general relativity), and a difference science for the micro, (atomic/quantum processes). The Galaxy Gravity Cycle (GGC) and the Atomic Model of Motion (AMM) bring these two separate sciences together for the first time. The Galaxy Gravity Cycle explain the atomic/quantum processes involved in creating the warped-space of General Relativity while the Atomic Model of Motion explain the atomic/quantum processes involved in the momentum, relativity, and gravity of masses.

Together, they bridge the quantum gap to explain how the atomic/quantum processes of momentum, relativity, and gravity (microphysics) operate within the cyclic flow of energy within a galaxy (macrophysics). The GGC and the AMM take us closer to what Einstein was seeking:

"For decades Einstein attempted to develop a unified field theory... connecting the movement of planets and stars with the operations of the tiniest subatomic particles." (Lacayo, 2014, 9)

The motion myth is the erroneous foundation upon which modern physics is built. That is why atomic/quantum physics had to break off of classical physics in order to progress in its development. Now that the motion myth has been exposed, the science of motion, with its atomic/quantum processes, can be further developed as a single, unified science.

In Conclusion

Even though the title of this book reads <u>Einstein Was Wrong!</u>, I really appreciate Albert Einstein's Relativity theories. Einstein was one thought away from being able to correct his theories and explain them from an atomic/quantum perspective. Had he realized that masses are always in motion, even when they appear to be at rest, he would have broken the curse of the great misperception. He would have been forced to look at the atomic/quantum causes regulating momentum, relativity, and gravity. Unfortunately, his two relativity theories still perpetuate the gap between our present, limited understanding of gravity and the atomic/quantum processes involved in sustaining gravity.

In the end, the purpose of science is not to compose reality to match our perceptions of it, but rather to change our perceptions until they conform to reality—to see things as they are even if it is different from how we want them to be.

Appendix A: Thesis

The Galaxy Gravity Cycle (GGC) and the Atomic Model of Motion (AMM) bridges the gap between the misrepresentation of space-time and the dice of quantum mechanics.

Appendix B: The Problem with Special Relativity

Does a Photon Have the Same Relativistic Qualities as Atom?

Quantum relativity as explained in this book as part of the Atomic Model of Motion (AMM) would be incomplete without explaining the problem with Einstein's Special Relativity. Einstein's dilemma was similar to the beanbag scenario. Replace the beanbag with a light pulse. Person A, on the moving train, sees a light pulse go four feet. Person B, standing on the stationary earth relative to the moving train, sees the same light pulse travel more than the four feet. (Please note that I understand the above example is not realistic and could not be observed with the natural eye. The scenario is based on the premise that if the distance the light pulse traveled was a lot longer, both observers would measure the same light pulse traveling a different distance.) This is because the light pulse will not only travel the four feet to the same target, but will also travel at an angle proportional to the speed of the train, just like the beanbag. Steven Hawking put it like this:

> Since the speed of the light is just the distance it has traveled divided by the time it has taken, different observers would measure different speeds for the light. In relativity, on the other hand, all observers *must* agree on how fast light travels. They still, however, do not agree on the distance the light has traveled, so they must therefore now also disagree over the time it has taken. (The time taken is the distance the light has traveled – which the observers do not agree on – divided by the light's speed – which they do agree on.) In other words, the theory of relativity put an end to the idea of absolute time! (Hawking, 1996, 21-22.)

This scenario only exists if you assume light has the same relativistic qualities as mass (atoms), meaning that mass (atoms) can obtain an apparent state of rest in any inertial frame. Without understanding the role of the atom in quantum relativity, one could easily make this mistake.

56

To answer the dilemma proposed by Stephen Hawking, I ask the following question: *What is the major difference between a beanbag (a collection of atoms) and a light pulse?* A beanbag (atoms) can appear to be at rest in any inertial frame. On the other hand, an emitted light pulse is never at rest in any inertial frame. This means that an emitted light pulse follows a different set of rules in how it moves through space than atoms in the form a beanbag. For example, it is common knowledge that the speed of light is unaffected by the speed of the mass (atoms) emitting it, but the speed of an object (atoms) is directly affected by the speed of the vehicle (more atoms) from which it is released. This major difference is due to the fact that in Galilean and Newtonian physics, objects (atoms) can acquire a perceived state of rest within an inertial frame—meaning synchronized momentum patterns—whereas a freely moving light pulse never finds rest within any inertial frame.

Let's quickly diagram the problem.

Illustration--3

Light Clocks

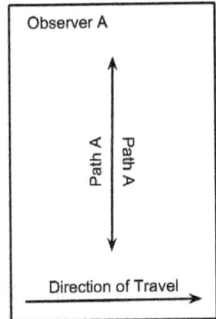 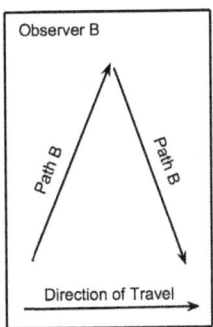

If *observer A* is traveling in a train with a hypothetical light clock, the light will move in an up and down motion in reference to *observer A* (Path A). *Observer B*, who is stationary to the motion of the train that is carrying *observer A* and the light clock, will observe the path of the light moving at angles in the direction of the uniform motion of *observer A* and the light clock (Path B). The distance of *path A* is different than the distance of *Path B*. Yet, from each person's perspective, they are both correct. This is quite a paradox if the speed of light is constant for all observers and yet travels two different distances. Einstein's idea was to take the different distances of travel and divide them by the same light speed; you end up with two different times for the same event. Mathematically, it makes sense.

Illustration—3 is a classic example used to explain time dilation of Special Relativity in many up-to-date encyclopedias. In illustration—3, if you replaced the light pulse with a bouncing ball, this would be an example of classic Galilean relativity for which I have already provided an explanation on a quantum level. However, a light pulse and a bouncing ball are not the same, so you shouldn't expect the same results. Einstein's mistake was to assume that a light pulse operated by the same rules as atoms in Galilean relativity.

Einstein failed to conclude that if the speed of light is independent of the motion of the mass emitting it (Alväger et al. 1964) then its direction or path should also be independent of that motion.

The **speed** and **direction** of emitted light pulses are unaffected by the motion or momentum of the objects emitting them. An object of mass, on the other hand, which can be at the same momentum level as the observed inertial frame within which it is contained, is directly affected by the speed of the object releasing it. Atoms, which make up objects and masses, must be treated differently than freely moving light pulses, which are never at the same momentum level of any observed inertial frame. In other words, freely moving light pulses cannot be regarded in the same manner as masses (atoms) in respect to Galilean relativity and inertial frames.

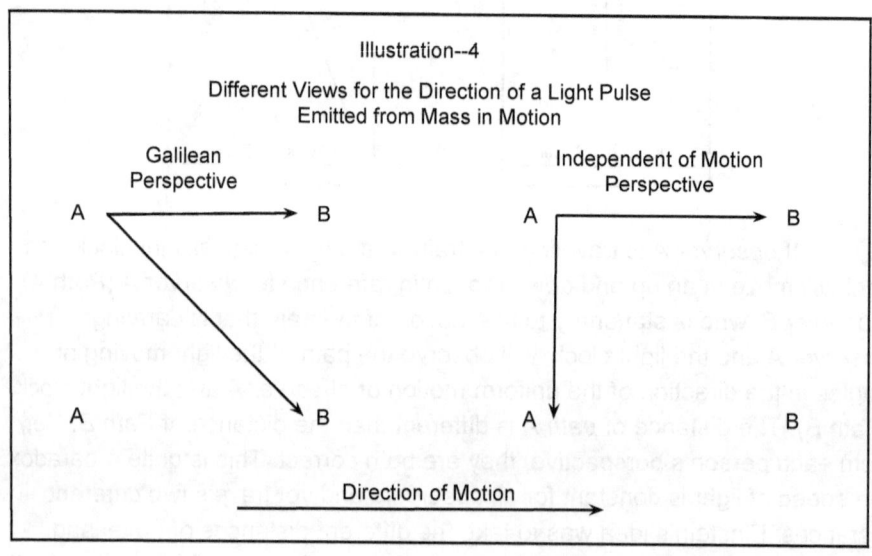

Illustration--4

Different Views for the Direction of a Light Pulse
Emitted from Mass in Motion

Galilean Perspective

Independent of Motion Perspective

A → B   A → B

A → B   A → B

Direction of Motion

In illustration—4, *A* represents uniformly moving mass in the direction of *B; A* also represents the emission of a light pulse perpendicular to the motion of the mass emitting it. The *Galilean Perspective* demonstrates light's direction or path of travel *dependent* on the motion of the mass emitting it—a continuation of Galilean relativity. The *Independent of Motion Perspective* demonstrates light's direction or path of travel independent of the motion of the mass emitting it.

Just as the **speed** of light is unaffected by the motion of the mass emitting it, the ***path or direction*** it travels after its emission should also be unaffected by the motion of the mass emitting it. This is because light does not have relativistic qualities, whether confined within mass or freely flowing in space. It goes back to comparing the motion of a photon to the motion of an atom. A photon has one fixed speed through space as calculated by Maxwell, whereas an atom's speed through space varies according to the momentum pattern and energy level of that atom. For this reason, when mass is pushed away from a moving body, its new momentum is a continuation of its previous momentum, whether dropping a package out of a flying airplane or throwing a baseball from a moving vehicle. On the contrary, the speed of light is unaffected by the speed of the object emitting it, such as turning a flashlight on from the same moving vehicle from which you threw the baseball.

I remember someone telling me how Special Relativity was demonstrated in a college class. One student took a piece of chalk and perpetually drew a line going up and down while staying in one place. The second person did the same up and down motions as the first person while walking along the chalkboard from one end to the other end. This supposedly demonstrated how light travels different distances in different inertial frames, creating the need for the contraction of time to account for the differences in distances. Unfortunately, this isn't how light works. A photon does not have the same relativistic quantities as an atom. This means that the speed of light is not the same for differing inertial frames (all moving bodies) as postulated by Einstein (who must have assumed that light had the same relativistic qualities as an atom), but rather, the speed of light operates independently of all inertial frames. Once light is emitted from the confined energies of inertial mass, its *speed* and *direction* of travel move independently from the momentum of the mass emitting it.

Of all people, why would Einstein make this crucial mistake and postulate that the speed of light is the same for all inertial frames? He was stuck in a motion myth paradigm, the idea that objects just mysteriously move through space. He failed to theorize the role of the atom in momentum, relativity, and gravity. For Einstein, because objects just mysteriously move through space, Galilean relativity remained a mystery. No one, including Einstein, could scientifically explain why the laws of physics are the same for all inertial frames, (quantum relativity). In his ignorance, Einstein assumed that if the laws of physics are the same for all inertial frames, then the fixed, unchanging speed of light as calculated by Maxwell must also be the same for all inertial frames. It was a terrible miscalculation on Einstein's part that reflected the scourge of the motion myth paradigm.

When Galilean relativity is explained from a quantum perspective using momentum patterns and energy levels, then one can easily see that a photon operates differently than an atom. An atom can obtain an apparent state of rest in differing inertial frames, whereas a photon never obtains an apparent state of rest in any inertial frame. Had Galileo been able to observe the energy levels of atoms, he would have noticed that they change to correspond with differing inertial frames, (AMM).

But alas, in 1905, Einstein's application of time (with the help of the Lorentz transformation) in his paper *On the Electrodynamics of Moving Bodies* delayed the discovery of the role of the atom in momentum, relativity, and gravity.

Appendix C: Einstein Continued...

The following appeared in a recent TIME Book celebrating Einstein's life. In referring to Einstein in the year 1929, "He optimistically told an English Newspaper that his new work proved at last that 'the force which moves electrons in their ellipses about the nuclei of atoms is the same force which moves our earth in its annual course about the sun, and is the same force which brings to us the light and heat which makes life possible upon this planet." The piece goes on to say that "...Einstein's colleagues in the scientific community were not impressed. For one thing, in order to arrive at his new theory, Einstein had violated rules established by his own general theory of relativity. Within a few years even he was obliged to admit that he had failed once more." (Lacayo, 2014, 78). Isn't it interesting that the scientists mentioned above appealed to general relativity as the basis or foundation from which to compare or contrast a new theory? The thesis of this book is that modern physics is built upon an erroneous foundation. General relativity is a part of that erroneous foundation.

But when one starts with a wrong premise, no amount of patching can right the problem. (Seifer 1996, 21)

In order to continue Einstein's quest for a unified field theory, we must first correct Newton's mistake and build from there. By doing away with the wrong premise—the motion myth paradigm—we can move forward on a better foundation. The Atomic Model of Motion (AMM) and the Galaxy Gravity Cycle (GGC) are only a starting point in the quest to further develop the science of motion.

Appendix D: What are Waves?

Physical waves are descriptions of atomic/quantum processes. Whether water waves, sound waves, or waves traveling through a rope, these types of physical waves are manifestations of the interaction of atoms of differing and changing momentum patterns and energy levels.

Quantum relativity helps us understand how energy moves through objects. As I try to explain energy to my sixth grade students by making waves with a rope, I ask them: *What is it that is traveling through the rope?* Someone will usually give the answer of energy. *Then I ask: What is energy? How does energy make the rope move?* This question of how energy moves through the rope as it is whipped to make wave motions perplexed me. A reaction takes place, but why, or even more important, how?

An in-depth look into quantum relativity explains the answer to this question. The observed movement of the rope is caused by a chain reaction on the atomic level. Before the rope is whipped, the momentum patterns of the atoms from which the rope is made are in par with the momentum patterns of the atoms of an orbiting earth through space. As the rope is whipped, the momentum patterns of the atoms making up that part of the rope change. The momentum patterns of those atoms react immediately to this disruption by absorbing or emitting energy to exactly correspond to the change. This disruption of momentum patterns disrupts the momentum patterns of corresponding atoms along the length of the rope. You end up with atoms in the same rope at varying momentum patterns and energy levels. In other words, atoms within the same rope temporarily experience different momentum patterns as they absorb and emit energy in an attempt to reestablish and maintain equilibrium.

The chain reaction of changing momentum patterns of the atoms making up the rope creates the resulting and observable movement of the rope in the form of a wave. Because the atoms of the rope are bonded together, the disruption of the momentum pattern of one atom causes a disruption to another and another, creating the chain reaction throughout the entire length of the rope. What is happening at the speed of light on the quantum level, the absorption and emission of energy accommodating the momentum changes of atoms, we see as a comparatively very slow moving wave through the entire length of the rope. Quantum gravity quickly returns the rope to its original state when the source of the wavelike motions cease.

When you drop a stone in water, it appears that energy moves through the water in the form of waves. What is really happening is the absorbed energy that caused the stone to accelerate disrupted the momentum patterns of the lower energy level atoms of the water. The waves in the water are the chain reactions of disrupted momentum patterns and their subsequent absorptions and emissions of energy to compensate for the changes. The waves are the result of changes in the momentum of the atoms that make up the water molecules. The energy didn't need water to travel through. The energy was the water at varying different energy levels. This created the waves. (From Newton's Mistake, a previous book by the author.)

Bibliography

Brian, Denis. *Einstein: A Life.* New York: John Wiley & Sons, Inc. 1996

Hawking, Stephen W. *A Brief History of Time.* New York: Bantam Books. 1988.

Hawking, Stephen W. *A Brief History of Time.* New York: Bantam Books. 1996.

Lacayo, Richard. Time Home Entertainment. *Albert Einstein: The Enduring Legacy of a Modern Genius.* New York: Time Books. 2014.

Seifer, Marc J. *Wizard: The Life and Times of Nikola Tesla: Biography of a Genius.* New Jersey: Carol Publishing Group. 1996.

www.ingramcontent.com/pod-product-compliance
Lightning Source LLC
Chambersburg PA
CBHW070941180526
45168CB00003B/1138